June 1985

Ashok

Thanks both for your early
assistance in the development of
the environmental auditing
practice and for your continuing
support and encouragement.

Jado
Gil
Maryanne

ENVIRONMENTAL AUDITING

ENVIRONMENTAL AUDITING

Fundamentals and Techniques

J. LADD GREENO
GILBERT S. HEDSTROM
MARYANNE DiBERTO
Center for Environmental Assurance
Arthur D. Little, Inc.

JOHN WILEY & SONS
New York · Chichester · Brisbane · Toronto · Singapore

Much of the material presented herein originally appeared, in slightly different form, in the Edison Electric Institute *Environmental Auding Workbook* prepared by Arthur D. Little, Inc., under contract to Edison Electric Institute, copyright © 1983 by Edison Electric Institute and used with permission.

This publication is designed to provide accurate and
authoritative information in regard to the subject
matter covered. It is sold with the understanding that
the publisher is not engaged in rendering legal, accounting,
or other professional service. If legal advice or other
expert assistance is required, the services of a competent
professional person should be sought. *From a Declaration
of Principles jointly adopted by a Committee of the
American Bar Association and a Committee of Publishers.*

Library of Congress Cataloging in Publication Data:

Greeno, J. Ladd.
 Environmental auditing.

 Includes index.
  1. Environmental auditing. I. Hedstrom,
Gilbert S. II. DiBerto, Maryanne. III. Title.
TD194.7.G74 1985   658.4'08   85-6248
ISBN 0-471-81984-0

Printed in the United States of America

10 9 8 7 6 5 4 3 2 1

FOREWORD

Environmental auditing has emerged over the past five years as a distinct specialty in corporate environmental programs. The environmental auditing programs currently in place have drawn heavily from two primary sources—technical investigations and internal auditing practice. Each current practitioner has developed his or her own special blend of procedures reflecting program objectives and company culture. The result is a wide diversity in practices—a very natural and healthy reflection of the current stage of development of environmental auditing as a distinct environmental discipline. Within this diversity, however, there are distinct principles emerging.

The environmental auditor, like counterparts in the internal audit and technical investigation fields, must communicate information that is at the same time complete, technically correct, and useful to management. Few mistakes will be allowed. To avoid mistakes, the environmental auditor must work within a carefully considered system of procedures designed to achieve program goals. The prudent auditor will draw on the growing body of experience in the field as a basis for both his or her original program design and regular self-evaluations.

To this point, the environmental auditor has had as a resource numerous references in the related fields, occasional technical seminars (of highly uneven quality), and personal communication among practitioners.

This, the first authoritative reference book on the subject, very competently captures the emerging principles and provides practical guides to program design. The authors have provided an important contribution to environmental auditing and those who practice it—both novice and experienced practitioners alike.

RALPH L. RHODES

Director
Health, Safety &
Environmental Surveillance
Allied Corporation

January 1985

PREFACE

Environmental auditing is a new and still evolving endeavor. Its roots date back to the early to mid-1970s when a handful of industrial companies, working independently and on their own initiatives, developed environmental audit programs as internal management tools to help review and evaluate the status of the company's operating units. Formal internal environmental audit programs in industry date back to 1972.

From the beginning, environmental auditing has been practiced in different ways by different companies. As described throughout this book, these differences in audit philosophy and approach still exist today.

As regulatory requirements and complexities continued to proliferate during the second half of the 1970s, the number of companies with environmental audit programs continued to grow. For example, in our nationwide survey of 110 of the larger industrial companies conducted in May 1979 by Opinion Research Corporation, 68% of the companies reported that they had an environmental auditing program in place.

As environmental auditing began to receive growing and more widespread attention within the private sector, it also became the subject of some interest among the regulators. One particularly early example of this regulatory interest dates back to October 1979 when the U.S. Environmental Protection Agency issued a draft report calling for independent, certified third-party environmental "auditors" who would visit plants, collect samples, perform analyses, and report results back to governmental authorities. While this governmental environmental auditing concept never made it beyond the draft report stage, it did receive considerable attention (and opposition) within industry.

In the meantime, voluntary environmental auditing programs within the private sector continued to grow, both in number of programs and in sophistication of audit approach. Seminars and conferences on the subject began to be held and became commonplace relatively soon. Similarly, articles and papers on environmental auditing began to appear more frequently in the business press.

Today, interest in environmental auditing continues to grow in both private and public sectors. Hence, the number of practitioners of environmental auditing and the roster of those who are helping to shape the area continue to expand. Furthermore, auditing will almost certainly continue to develop and evolve.

Despite the strides made, environmental auditing still has all the hallmarks of an embryonic endeavor. Many companies are evaluating the need to establish an environmental audit program. Those that do have programs practice a wide variety of approaches. They have not even settled on a standard name to describe these programs. They are referred to as *review, surveillance, survey, assessment, appraisal, evaluation*, and *audit*, to mention a few. (Because *audit* is the most common, we use that term throughout the book.)

This book has been written to provide management and environmental staff with the information and background to decide whether an audit program makes sense for their organizations and, for those where an audit program is deemed appropriate, the information needed to develop an effective environmental audit program tailored to individual organizational requirements.

This book is for the established practitioner and novice alike. It has been designed to help familiarize those interested in environmental auditing with the basic concepts, techniques, and methodologies of the field.

The book contains many examples and specifics of alternative approaches for those readers whose interest extends beyond general familiarization. Individuals developing new environmental audit programs will find both a discussion of the major alternatives and key design variables and a specially designed workbook section helpful in identifying and specifying alternatives appropriate to their organizations. Lastly, experienced environmental auditors will find information useful in reviewing their current programs and considering modifications and revisions that could enhance the effectiveness of their programs.

The book has been organized into five major parts:

Part 1 provides an introduction to environmental auditing. Introductory material and background information are presented in the chapters "What is Environmental Auditing?," "Why Audit?," and "What Does an Audit Involve?"

Part 2 describes the basic design issues facing those charged with the responsibility for developing an environmental auditing program for their organization. Collectively, the chapters on goals, scope, organization and staffing, record keeping and reporting, and action planning and follow-up provide a discussion of the basic framework of an environmental auditing program. Within each chapter, the major alternative approaches are presented and discussed in terms of their general strengths and limitations. Case examples illustrating the alternatives and presenting current practices are provided in each chapter.

Part 3 deals with the techniques of conducting an environmental audit. Audit planning and other pre-audit activities are discussed as are audit protocols and questionnaires, the basic tools that guide the environmental auditor in conducting the audit. The chapter "Field Work" details the basic process used by environmental auditors to gather, develop, and review information; and presents various techniques to increase the effectiveness and efficiency of the audit effort. Audit note taking and documentation techniques are discussed in the chapter "Working Papers and Audit Record Keeping." Approaches and techniques for reporting audit findings and results are explored in detail in the final two chapters of Part 3.

Part 4 serves as a guide to the manager charged with establishing or administering an environmental audit program. This section of the book combines text with specially designed worksheets to provide a structured review of some of the important aspects of program design and implementation: overall strategy for program design and implementation, start-up concerns, and selecting facilities for the initial audits. Program implementation aspects featured in the interactive workbook format of Part 4 include strategy formulation, start-up considerations, and selecting facilities for the initial audits. Also discussed in Part 4 are the ongoing program administration issues of developing a schedule and audit cycle and records retention policy. Lastly, Part 4 presents a guide for managing audit quality and critiquing established audit programs.

Part 5 identifies briefly emerging trends and forecasts likely future directions for environmental auditing in terms of implications for today's programs.

Finally, the Appendixes present a set of sample environmental audit protocols and questionnaires to serve as basic models or starting points that can be modified or tailored to suit individual company needs.

J. LADD GREENO
GILBERT S. HEDSTROM
MARYANNE DIBERTO

Cambridge, Massachusetts
March 1985

ACKNOWLEDGMENTS

This book owes its existence to a number of people and organizations whose inspirations, ideas, experiences, efforts, and support have helped to make it a reality. We are indebted to our environmental auditing clients who provided the experiences that serve as the basis for this book. We are also grateful to the many practitioners in this emerging field who have shared their experiences and ideas with us. We also wish to thank the many companies whose participation in our environmental auditing surveys has not only furthered our understanding of the many approaches and techniques in use but also helped us refine our thinking and shape many of the concepts presented in this book.

We want to pay special tribute to Edison Electric Institute and its Environmental Auditing Educational Task Force for the opportunity they afforded us through preparation of the Edison Electric Institute *Environmental Auditing Workbook* to pull together the many papers, memos, and thoughts developed over the course of seven years.

Lastly, we are deeply indebted to our many colleagues at Arthur D. Little, Inc. for their help and support. Melanie Getchell, Janet Galligani, and Karen Keatley skillfully transferred our often illegible scribbles into a series of manuscripts. Lou Visco helped sharpen our fuzzy thinking and corrected our poor grammar while editing our drafts. We owe a special debt of gratitude to these four individuals. They brought order to a chaotic process, helped to keep us on schedule, and used their superb skills to make up for our delays. The task would have been impossible except for their willing and capable assistance.

We especially wish to express our thanks and gratitude to John Funkhouser. John founded the Center for Environmental Assurance, a special unit at Arthur D. Little dedicated to advancing the state of the art of environmental, health, and safety management, that provides a focal point for our firm's work in environmental auditing and the home base for our work. John provided each of us with our start in environmental auditing; led many of the client assignments that provide the foundation for the book; fathered many of the concepts presented in this book; and has been a constant source of ideas, advice, and support. His enthusiasm for the book was invaluable and his encouragement provided the motivation needed to go forth and split infinitives.

J.L.G.
G.S.H.
M.A.D.

CONTENTS

ENVIRONMENTAL AUDITING

PART ONE

INTRODUCTION TO ENVIRONMENTAL AUDITING

CHAPTER ONE

WHAT IS ENVIRONMENTAL AUDITING?

BACKGROUND

Environmental auditing, the process of determining whether all or selected levels of an organization are in compliance with regulatory requirements and internal policies and standards, has proven to be a powerful component of environmental management programs. Today, many companies are devoting a sizable and increasing proportion of capital expenditures, operating costs, and managerial energies to programs for pollution control, worker health and safety, product safety, and loss prevention. In addition to using sophisticated environmental control technologies, many of these corporations are also devoting considerable attention to environmental auditing as one way of formalizing some aspects of their environmental, health, and safety management systems.

Auditing, in general, is a methodical examination—involving analyses, tests, and confirmations—of local procedures and practices whose goal is to verify whether they comply with legal requirements, internal policies, and accepted practices. Auditing differs from assessment in that it requires collection and documentation of competent and sufficient evidence rather than an opinion based primarily on professional judgment.

The concept of environmental auditing is still relatively new. Consequently, there are still many interpretations of exactly what is meant by the term. Chapter 1 describes the nature of environmental auditing and discusses the role of auditing in the overall environmental, health, and safety management systems.

THE NATURE OF ENVIRONMENTAL AUDITING

Whatever the name used to describe an environmental auditing program—audit, review, surveillance, survey, assessment, evaluation, or appraisal—the key point is that such programs audit or review the environmental status of individual facilities.

In some respects, environmental auditing techniques have borrowed from those used in financial auditing. In financial auditing the basic functions of the auditors are to verify the adherence to standards, to certify that accounting procedures are appropriate for those standards, and to certify the accuracy of corporate financial records. Many environmental auditing programs employ verification techniques patterned after financial auditing for confirming compliance with federal, state, and local regulations; assessing the adequacy of environmental management programs to ensure that policies and procedures are followed and that regulations are adhered to; and verifying the validity of environmental records and reports.

One key difference between environmental audits and other types of audits is the existence or absence of standards. Few standards exist for environmental auditing. Financial audits have standards promulgated by the Financial Accounting Standards Board, and Food and Drug Administration audits use Good Manufacturing Practices. Another difference is the number of systems in place. The detailed, coordinated financial accounting systems that are in place can be subjected to a financial audit. However, aside from such things as pollution control data, consent agreements, and memoranda of understanding, relatively little environmental information is typically in place to be audited. Table 1-1 provides a comparison of different types of auditing.

AUDITING AS A COMPONENT OF ENVIRONMENTAL MANAGEMENT

An *environmental management system* is the framework for or method of guiding an organization to achieve and sustain performance in accor-

TABLE 1-1. COMPARISON OF AUDITABILITY

	Financial Auditing	Pollution Control Auditing	Occupational Safety Auditing	Occupational Health Auditing	Product Safety Auditing
Basic purpose	To protect stockholders by showing that proper procedures are being employed to account for financial transactions and presume corporate assets	To provide assurance to corporate management that environment and public are protected against acute or chronic hazards	To ensure corporate management that workers are protected from being hurt or killed in an accident or gradually harmed as a result of long-term adverse working conditions	To assure corporate management that workers are protected from acute or chronic health hazards	To provide assurance to corporate management that consumers are protected from being hurt or killed by the purchase or use of a product
Group at risk/ interested parties	Board of directors Stockholders Corporate management Investment community Public	Board of directors Neighbors General public Environment Consumers	Board of directors Employees and unions Families of employees	Board of directors Employees Families of employees	Board of directors Consumers Public interest groups Public interest
Nature of significant surprises	Misrepresentation of net worth, profits, assets, inventory, or reserves Unaccountable funds Embezzlement	Damage to the environment Contamination of food chain Chronic or acute health effects	Injury or death of worker(s)	Illness or death of worker(s)	Injury or death of consumer(s)
Current status of auditing	Firmly established Widely accepted Separate industry Generally accepted auditing principles and standards Detailed procedures	Evolving Growing acceptance of auditing concept Few auditing standards or principles Considerable diversity of auditing practices	Evolving Few auditing principles or standards Long history of safety inspections Good records generally available	Embryonic Environmental auditing principles or standards Sometimes combined with occupational safety auditing	Embryonic Considerable interest, however, relative diversity between auditing programs Few auditing principles or standards

dance with established goals and in response to constantly changing regulations, social, financial, economic, and competitive pressures, and environmental risks. When operating effectively, a corporate environmental management system provides management and the board of directors with the knowledge that

The corporation is in compliance with federal, state, and local environmental laws and regulations.

Policies and procedures are clearly defined and promulgated throughout the organization.

Corporate risks resulting from environmental risks are being acknowledged and brought under control.

The company has the right resources and staff for environmental work, is applying those resources, and is in control of its future.

Both the formality and complexity of environmental management systems can and do vary tremendously. Generally speaking, the more critical to the organization a desired action or outcome becomes, the more appropriate or desirable a formal environmental management system may be. Regardless of formality, however, most environmental management systems consist of several interrelated functions.

Planning.   Setting goals, establishing policies, defining procedures, and establishing program budgets.

Organizing.   Establishing the organizational structure, delineating roles and responsibilities, creating position descriptions, establishing position qualifications, and training staff.

Guiding and Directing.   Coordinating, motivating, setting priorities, developing performance standards, delegating, and managing change.

Communicating.   Developing and implementing effective communications channels within corporate, within the division, and with external groups, including regulators, as appropriate.

Controlling and Reviewing.   Measuring results, acknowledging performance, diagnosing problems, taking corrective action, and purposely seeking ways to learn from past mistakes and thereby creating improvements in the system.

Within the context of the overall environmental management system, the review function can provide a basis for guiding, measuring, and evaluating performance. The purpose of this function is to evaluate the extent to which a company's operations are carried out in a manner consistent with and supportive of applicable regulations and corporate policy. Environmental auditing is one *part* of the control and review function, and thus a small, but by no means insignificant, part of the overall environmental management system.

BASIC ENVIRONMENTAL MANAGEMENT PHILOSOPHIES

The purpose of an environmental auditing program varies considerably with the overall philosophy or thrust of environmental management within a corporation. The management of environmental activities in U.S. corporations can be described in terms of three stages of evolution. One facet of this progression (described below) is the broadening of the scope of problems the company chooses to address.

Stage I: Problem Solving

In Stage I, a company's environmental management efforts can be characterized by the desire to "stay out of trouble." The principal focus is on solving the immediate and most recognized environmental problems and on avoiding unnecessary costs resulting from increased staff or capital expenditures. Here, environmental management systems tend to be nonformal, and the responsibility for environmental management, for the most part, lies with lawyers, engineers, and other specialists who tend to focus on specific plant problems and concerns. They tend to address only the "necessary" laws and regulations—those that leave no room for interpretation—and the most significant hazards.

Few Stage I companies have a formal environmental audit effort. They frequently see little need to look for problems and often feel that an audit might simply add to costs. Where audit programs do exist in Stage I companies, they focus on assessing whether any major problems are present.

Stage II: Managing for Compliance

In Stage II, a company builds a more formal system to manage for the desired level or degree of compliance. This shift may stem from management's desire to better manage what is prescribed by the law or company policies and procedures. The principal focus of the environmental, health, and safety management system is on achieving and maintaining the desired level of compliance with various regulatory requirements. Here, environmental audit programs tend to include not only an assessment of problems (and perhaps of good practices), but also determination and/or verification that compliance is being achieved.

Stage III: Managing for Environmental Assurance

In Stage III, the basic management philosophy is that the full range of potential environmental risks to the corporation and to the environment (not just those covered by the current regulatory framework) must be managed. Said another way, not only are compliance-related risks important to the corporation, but also other risks not yet adequately covered by regulatory requirements or existing external standards are important. The principal focus is on building an environmental management system that emphasizes protecting internal resources and the external environment from injury by looking for and anticipating hazards as well as by managing the consequent risks. In Stage III companies, the environmental audit program often assesses the appropriateness of the environmental management system and verifies its effectiveness in addition to assessing problems and verifying compliance.

AUDITING IN THE CONTEXT OF ENVIRONMENTAL HAZARDS

Whether the overall corporate philosophy is Stage I, II, or III, there is a critical link between the audit program and the complexities of environmental risk.

To define this link, it is useful to distinguish among the types of environmental hazards. One approach is to identify the causes of hazardous industrial conditions.

People who do not fully understand regulations and procedures, do not pay attention to details, and so on.

Physical facilities that are inadequately designed, poorly maintained, inadequately protected, and so on.

Management systems that are limited in scope and are inflexible, or not supportive of open communication, and do not clearly delineate authority, responsibility, and accountability.

Procedures that are inadequate, inappropriate, or outdated.

External forces, such as earthquakes, aircraft, storms, riots, and sabotage.

Competing internal pressures, such as making profits and maintaining manufacturing output.

Environmental hazards can also be viewed from the perspective of known versus unknown hazards. Figure 1-1 depicts schematically the fact that we understand only a portion of the hazards we could know about based on present knowledge. The first category, labeled A, includes those hazards for which standards have been established and the corporation can actively managing against. Category B represents those hazards that are known and can be managed, but for which explicit standards do not exist. Category C includes those hazards not yet identified by a given company, but which could be identified by applying the right resources and expertise. Category D recognizes that some hazards exist that none of us know about. Experience tells us that in the future we will be discovering more and more hazards that are simply unknown today.

A basic goal of environmental management (assuming reasonable standards) is to constantly enlarge area A in relation to areas B and C, and area B in relation to area C. Note that, from an auditing perspective, area A can include not only regulatory standards, but also company, division, or facility standards (or standard procedures).

Generally speaking, environmental auditing becomes less effective as you move from the center of Figure 1-1 outward. Where there are established standards—be they performance standards, design standards, technology standards, or management standards—an audit can provide a methodical comparison of current practices (or current results) against those standards. Where hazards are recognized yet no standards exist (area B), an audit can assess, but not verify, the appropriateness of the management approach. With respect to area C, the basic objective is to identify knowable but previously unidentified hazards. Other approaches may be better suited to identification of an unknown hazard. Nevertheless, environmental audits should help identify and eliminate internal blind spots and communication gaps where hazards are generally acknowledged by the company but are unknown or unrecognized by the facility. Lastly, it is unlikely that an environmental audit would serve to identify unknown hazards.

Although industry's concern with protecting human health and reducing environmental impacts has been growing rapidly in recent years, information and understanding about environmental matters are still relatively limited. Our intuitive sense of what is needed to protect the environment frequently exceeds by a considerable amount our data base for logical, orderly decision making. Therefore, decisions about the management of environmental hazards need to be based on a process that

FIGURE 1-1. Conceptual range of environmental hazards.

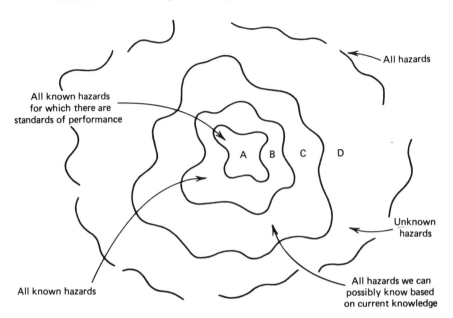

both allows for intuitive judgments based on imprecisely derived values, assumptions, and principles, and strives to be as explicit and precise as the data will allow—always seeking substantiation of the original intuitive judgments.

It is in the context of this increasingly complex arena that environmental audit programs have begun to play a powerful role. Chapter 2 explores the driving forces behind the development of environmental audit programs.

CHAPTER 2

WHY AUDIT?

Environmental audit programs do more than verify environmental compliance. They can have a wide variety of objectives and benefits depending on a corporation's culture, management philosophy, and size, and the specific needs of the individual within the corporation who is the driving force behind the establishment of the environmental auditing program. Furthermore, audit programs differ by the motivations behind the program, by the role the audit program plays in the corporation's overall approach to environmental management, and by the level of involvement of different functions within the corporation. Chapter 2 discusses motivating factors for setting up a program, various audit program objectives, and commonly perceived benefits to industry of environmental auditing. Chapter 4 of this book focuses on the key issues to consider when selecting from among these various program objectives.

MOTIVATING FACTORS

The impetus for establishing an environmental audit program can come from different people and can arise for a variety of reasons. Many programs have been motivated by the desire of the board of directors or chief executive officer (CEO) to obtain assurance and comfort that the corporation is in charge of and adequately handling its environmental responsibilities. On the other hand, programs have also developed from initiatives on the part of lower or middle management to improve their existing environmental management activities and to keep up with what other companies are doing.

Some environmental audit programs have been motivated by the occurrence of an environmental problem or incident. Others have been established in response to a desire to anticipate and head off potential problems.

PROGRAM OBJECTIVES

Just as a number of factors can lead a corporation to establish an environmental audit program, there also are a variety of audit program objectives. The following objectives are not mutually exclusive. The relative importance given to them, however, will influence the shape of the auditing program. For example, where the focus is to provide assurance, the program typically is supported by top management—the report may go to the board of directors or CEO and there often is a scheduled reporting basis. By contrast, when the emphasis is on helping the facility manager, program support is typically at a lower level within the organization—the report may go to a vice-president, yet there is often no scheduled reporting basis.

Virtually all environmental audit programs are to some degree, compliance oriented—established to identify and document the environmental compliance status of operating facilities. A compliance-oriented audit program can provide a systematic check on the extent to which a facility complies with the terms of its various environmental operating permits. Or, it can review the extent to which a facility complies with a complex set of federal, state, and local regulatory requirements. Such a program can also detect environmental hazards that are both known and regulated (see area A in Figure 1-1), but does not necessarily detect the hazards beyond those for which standards currently exist (areas B, C, and D in Figure 1-1).

Within the overall objective of identifying and documenting compliance status, a program may take a variety of forms, such as, identifying and documenting compliance discrepancies and recommending steps to facilitate corrective action; helping facility management understand and interpret regulatory requirements, company policies, and guidelines; and identifying differences among or shortcomings at individual facilities, or patterns of deficiencies that may emerge over time. Audit programs may also differ in the time frame under consideration—some companies focus on compliance over time, while others focus on compliance at the time of the on-site review.

With the determination of compliance status as a basic program element, audit programs are established to achieve different objectives. Some programs are explicitly established to provide assurance to top management that the facilities are in compliance, while others specifically focus on helping the facility manager understand and conform to regulatory requirements and corporate policies. Some programs place special emphasis on assessing risks while others place equal emphasis on identifying ways to optimize resources. In short, companies answer the question Why audit? in many ways.

Provide Assurance to Management

Some environmental audit programs are designed to assure management—often top management—that the company's potential exposure to regulatory compliance problems is at, or is being reduced to, an acceptable level. Assurance can also be provided by confirming that systems are in place and that the facility's operations are consistent with good practice. Such programs may be established to assure management that the company's institutional and ethical, as well as legal, responsibilities are being met.

Providing management assurance usually requires a determination of the facility's compliance status and an effective means of reporting that information to management. Such programs typically require a significant degree of independence on the part of the audit team from the facility being audited. In addition, programs established to serve the needs of top management generally demand a more rigorous and in-depth review of facility operations than do programs designed to serve the needs of facility management.

The impetus for programs designed to provide assurance to management typically comes from members of the board of directors or top management, who believe that corporate environmental protection is both important and necessary. Since the board of directors is charged with protecting the interests of the public and the company's stockholders, an environmental audit program can serve as a control and check to assure them that, to the extent possible, risks have been identified. Moreover, top management must manage the corporation according to the prescribed policy of the board of directors. An environmental audit program that is integrated throughout all levels of the organization can help top management know and be assured that the facility's operations are consistent with corporate policy. Such a program can further demonstrate corporate commitment to identifying potential environmental problems and taking reasonable steps toward resolving those problems.

Assist Facility Manager

Programs designed to assist facility managers are generally established not only to determine the facility's compliance status, but also to provide information that helps managers improve their performance and understand new or not widely known environmental requirements or corporate policies. Such assistance can be in the areas of understanding and interpreting regulatory requirements, identifying compliance problems, defining cost-effective measures that should be taken to achieve compliance, and identifying additional facility personnel needs and opportunities.

Another area of assistance is through employee training and awareness. A training component in an environmental audit program can help facility employees know and follow environmentally acceptable procedures as well as understand the environmental management practices and reduce human error in routine procedures and practices.

Assess Risks

Audit programs designed to assess risks go well beyond compliance-oriented programs in both focus and breadth. These programs are not confined to auditing facility performance against known standards (e.g., regulations); they may include examining potential hazards for which standards do not currently exist. Risk assessment programs may identify conditions at the facility that may have an adverse impact on the corporation, assess risks associated with those hazardous conditions identified, and determine what actions are necessary to control those risks.

Companies whose audit programs are designed to help manage risks place a strong emphasis on verifying that management systems are in place to identify and assess environmental hazards and risks. For example, in looking at the elements of a polychlorinated biphenyls (PCBs) management system, an auditor may want to examine the system or procedures in place for handling, storage, marking, cleaning up spills, inspections, record keeping, and an annual inventory. Some companies further expand this objective to look for risks that are not yet identified and focus on the need to reduce future risks. Auditing can thus be used to verify the process by which the company manages or seeks to identify unregulated or soon-to-be-discovered risks. (A discussion of auditing in the context of environmental hazards is provided in Chapter 1.)

Optimize Resources

Some audit programs are designed to optimize environmental resources in terms of both personnel roles and responsibilities and capital expenditures. Such programs, in addition to determining compliance, may identify current and anticipated environmental costs, recommend ways of reducing those costs, and identify potential longer-term savings.

Audit programs that are aimed at optimizing resources tend to focus on environmental cost savings and other economies available to a corporation. For example, this type of audit program may help to define facility personnel roles for specific environmental disciplines and those responsibilities necessary for carrying out those roles. It can define gaps in job responsibilities—where no responsibilities have been assigned or assignments have not been communicated appropriately. Similarly, such programs can identify efficient and cost-effective means of achieving compliance—by identifying, for example, less costly yet more efficient pollution control equipment, recommending less stringent permits for a facility to comply with applicable regulations, and, potentially, reducing current permit requirements. After seeing how other facilities handle certain problems, the audit team can share that information with facility personnel.

Other Considerations

In addition to the primary objectives for the audit program, a number of other considerations can influence the decision to establish an environmental audit program. These include the Securities and Exchange Commission (SEC) environmental disclosure requirements; environmental impairment liability insurance; environmental considerations relating to mergers, acquisitions, and divestitures; and the extent of environmental audit programs at peer companies.

SEC Requirements.   Federal securities laws do *not* require publicly held companies to conduct environmental audits. These laws, however, do require publicly held companies to disclose all "material" effects resulting from compliance with environmental statutes. (Material has been defined as affecting the competitive position of the firm, involving a claim of at least $100,000, or exceeding 10% of the current assets of the corporation.) The Securities and Exchange Commission requires disclosure of all material effects in regular common filings and in special filings such as proxy statements associated with pending mergers and prospectuses for bond issues. The SEC has, on occasion, demonstrated its intention to enforce the environmental disclosure requirements through injunctive actions and administrative proceedings.

Most companies already have well-established procedures for preparing disclosure documents required by the SEC. Generally speaking, environmental disclosures are handled by the basic process the company has established for identifying and preparing SEC reports of other (non-environmental) potentially material effects on the corporation. Environmental management, and in some instances the environmental audit program, may be asked to play a role in this process. However, because the SEC requirement applies only to fairly substantial and costly items, many companies have not found it necessary to formally tie the audit program into their SEC reporting process. Given the "double check on the system" nature of most environmental audits, most exceptions noted during the audit tend to be considerably less significant than the SEC reporting requirements. Furthermore, should a material deficiency turn up in an audit, it would be promptly reported to management via the established environmental auditing process. Thus, those who would need to know of the situation for SEC disclosure would have the needed information.

Mergers, Acquisitions, Financing, and Divestitures.   A flurry of activity in the buying and selling of facilities, operations, businesses, and even whole corporations—often without much understanding of or consideration to environmental issues of the company being purchased or sold—is focusing attention on the need for and value of better environmental information before negotiations are completed and the deal is consummated. Many companies with established environmental audit programs now routinely audit facilities prior to divestiture. Similarly, a smaller, but growing, number of companies are looking for audits of the facilities they are about to acquire—audits conducted by either their own environmental staff, the seller, or an independent, third-party environmental auditor. Although environmental audits are not required, full disclosure requirements similar to those of the SEC are often required for financing packages (such as bond issues).

Environmental Impairment Liability Insurance.   Tradition-ally, pollution liability insurance coverage for "sudden" or accidental incidents was offered as a part of a basic comprehensive general liability policy. Today, this coverage can be augmented by a new form of insur-ance policy, environmental impairment liability insurance, which covers nonsudden pollution incidents. This gradual-pollution insurance cover-age is still in its infancy. Nevertheless, it can have some important effects for deciding whether to establish an environmental audit program.

Several of the relatively few insurance carriers that offer nonsudden pollution coverage require an outside "audit" of sorts to be conducted to assist them in deciding whether to underwrite the environmental risks associated with the company that is applying for the coverage. These insurance audits have differed from many of the individual internal environmental audit programs both in the nature of the audit examination and the level of effort that is expended. Insurance audits often take a less detailed look than many of the corporate environmental audits. The level of effort often is less than a company's own environmental audit program.

As both environmental auditing and environmental impairment liabil-ity insurance continue to develop, there are likely to be more interfaces and greater linkages between the two areas. In fact, some companies seeking impairment insurance are already negotiating for consideration, at least in part, of their own environmental auditing program in lieu of the insurance carrier's audit.

Audit Practices at Peer Companies.   Another consideration in deciding whether to establish an environmental audit program is the extent to which peer companies have established audit programs. As the popularity of environmental auditing continues to grow, companies can feel a sort of "peer pressure" to establish programs comparable to those of their peers and consistent with developments within their industry segments.

COSTS AND BENEFITS OF AUDITING

In answering the question Why audit?, a company needs to look at both the costs involved and the benefits gained. The costs of an audit program include both direct and hidden costs. The direct costs include the salary and benefits of the program staff, any allocated expenses from other corporate staff departments, all related travel and living expenses, and other costs that are directly attributed to the program. The hidden costs include the time plant personnel devote before, during, and after the audit and the time others such as legal department or plant staff who may serve as audit team members, and perhaps engineering, contribute. Thus, the costs associated with the audit program need to be characterized just as they would for any other program and balanced with the benefits.

The benefits to industry of environmental auditing are as broad and varied as the audit program objectives discussed previously in this chapter. To distinguish between the two terms, we define an objective of an audit program as an end toward which effort is directed over a period of time. A benefit to a company of environmental auditing is an aspect or output of auditing that contributes significantly and positively to the achievement of corporate objectives. In short, companies work toward achieving audit program objectives so that they can gain the resulting value or benefit.

This distinction is important because the objectives of audit programs are commonly used in industry as the benefits per se. For example, a typical audit program objective is to determine the compliance status of individual facilities. The benefit is really not identification and documentation of compliance status, but rather may be increased environmental management effectiveness as measured by such things as an improved compliance record, reduced occupational hazards, fewer legal actions brought against the company, and so on. In this example, identification and documentation of compliance status are an objective of implementing the audit program, not the benefit.

What Are the Benefits?

The most significant benefits to industry of environmental auditing can be classified into two broad categories: increased management effectiveness and a feeling of increased comfort or security.

1. *Increased Environmental Management Effectiveness*. A significant number of the audit program objectives discussed above contribute toward increasing the overall environmental management effectiveness of the corporation. This increased effectiveness results from identifying and reducing "blind spots" that may exist, clarifying issues that might otherwise be interpreted differently at different facilities, and developing a more uniform approach to managing environmental activities through sharing information with and learning from other facilities.

To some extent, increased environmental management effectiveness can be quantitatively measured over time. Some basic measures include an improved compliance record, reduced number and size of fines, improved incident and accident statistics, reduced volume and size of legal actions, and reduced volume of environmental hazards. However, there are likely to be significant limitations to using such measures to evaluate management effectiveness. Some measures of improved environmental management effectiveness, such as an improved reputation or favorable publicity, are generally not quantifiable.

2. *Feeling of Increased Comfort or Security.* Many environmental auditing programs are established at the request of top management for the purpose of identifying and documenting the compliance status of individual facilities. The primary benefit of such programs is to provide top management with a sense of increased comfort or security that the company's potential exposure to regulatory compliance problems is being reduced. (This benefit is commonly perceived more by top management than by facility management.) The feeling of comfort is generally nonquantifiable and stems from the knowledge that operations are consistent with good practice, that control systems are in place and operating, and that legal and ethical responsibilities are being met.

Who Benefits?

Ultimately, the benefits to a company of environmental auditing are defined by a combination of the corporate benefits and the net sum of the benefits to various individuals or groups. An example of a benefit to the corporate entity might be improved company reputation. On the other hand, a benefit to a senior officer might be a feeling of increased comfort or security. The key individuals potentially benefited by an environmental audit program are listed below.

Stockholders
Board of directors
Corporate officers
Corporate environmental management
Legal department
Manufacturing manager
Facility manager,
Facility environmental staff

Line supervisors

Hourly workers

Each of the individuals listed above has a different measure of value so, for example, what is perceived as a benefit to a corporate officer is generally different from what a facility manager would call a benefit. Others who might be added to the list include customers, neighbors, and other members of the general public who may potentially benefit from an environmental auditing program.

TABLE 2-1.   EXAMPLE MEASURES OF AUDIT PROGRAM BENEFITS

Directly Influenced by Audit Program	Generally Quantifiable	Generally Not Quantifiable
Improved compliance record	√	
Reduced legal actions brought against company and/or individuals	√	
Reduced fines	√	
Improved incident/accident statistics	√	
Reduced volume of environmental hazards	√	
Improved worker health	√	
Improved reputation		√
Favorable publicity		√
Improved regulatory relations		√
Indirectly Influenced by Audit Program		
Decrease in business interruptions by identifying problems that could affect production	√	
Increased worker productivity from reduced environmental risks	√	
Reduced insurance rate	√	
Increased involvement in day-to-day environmental activities		√
Increased job satisfaction		√
Knowledge of job performance measures		√
Feeling supported by management		√

Source.   This table is based on "Benefits to Industry of Environmental Auditing," a report to the U.S. Environmental Protection Agency by Arthur D. Little, Inc. (EPA Report No. EPA-230-08-83-005).

How Do You Measure the Benefits?

While the perceived benefits of environmental auditing vary both among companies and among different individuals within a corporation, it is possible—to a limited extent—to measure the benefits of auditing. Table 2-1 provides some example measures that might be used to define audit program benefits. Table 2-1 distinguishes between those measures that are commonly considered to be directly influenced by the audit program and those not directly influenced. In addition, Table 2-1 categorizes each measure as being either generally quantifiable or generally nonquantifiable. In looking at Table 2-1, however, remember that many other factors typically contribute to achieving these benefits. Many of the objectives audit programs strive to meet are also objectives of other environmental management activities.

Summary

To sum up, the benefits to industry of environmental auditing are as broad and varied as the underlying reasons for which companies develop environmental auditing programs. There are a variety of risks and opportunities associated with environmental auditing as will be discussed throughout this book. In one sense, the clearest statement of audit program benefits is that those companies with established environmental audit programs feel strongly that the benefits are substantial. As a result, the number of companies with environmental audit programs continues to increase.

CHAPTER 3

WHAT DOES AN AUDIT INVOLVE?

Despite the variety of audit program objectives discussed in Chapter 2, some elements are common to most audit programs. Each audit program involves having a team of individuals conduct a field assessment, gather information, analyze information, make judgments about the facility's environmental compliance status, and report audit findings. Chapter 3 presents a brief overview of the key steps common to environmental audits. Each is discussed in detail in Chapters 7 through 14. While not all audit programs are designed precisely as outlined below, each makes some provision for including these steps in their audit process. What does vary among companies is the time and effort devoted to pre-audit, audit, and post-audit activities.

PRE-AUDIT ACTIVITIES

The environmental audit process actually begins with a number of activities before the actual on-site audit takes place. These activities include the selection of the facilities to be audited, the schedule of the facilities to be audited, the selection of the audit team, and the development of an audit plan which includes defining the scope of the audit, selecting priority topics to include, modifying the audit protocols, and allocating audit team resources. They may also include an advance visit to the facility to gather background information and/or administer questionnaires. Chapter 9 provides a more in-depth discussion of pre-audit activities.

KEY ON-SITE ACTIVITIES

The actual on-site audit typically includes five basic steps. (See Figure 3-1.)

   1. *Understanding Internal Management Systems and Procedures.* The first step that the audit team takes is to develop an accurate understanding of the facility's internal environmental, health, and safety management system—the set of formal and informal actions taken by the facility to assist in regulating and directing its activities that can impact the environment. Internal controls, in its broadest sense, refers to both the management procedures and the equipment or engineered controls that affect environmental, health, or safety performance. The auditor's understanding is usually gathered from multiple sources such as staff discussions, questionnaires, plant tours, and, in some cases, a limited amount of verification testing conducted to help confirm the auditor's initial understanding. The auditor usually records his or her understanding in a flow chart, narrative description, or some combination of the two in order to have a written description against which to audit.

   It is important that an environmental auditor not take too narrow a view of what is or is not an internal management control system. The basic goal in this step is to understand the various ways in which the facility intends that its environmental concerns be managed. In almost all organizations, many aspects of the facility's internal environmental management systems [such as a complete description of the program in place for National Pollutant Discharge Elimination System (NPDES) sampling, analysis, monitoring, and reporting] will not be documented or described in writing. However, selected management systems [such as a Spill Prevention Control and Countermeasure Plan (SPCC)] may be documented in enough detail to provide both an understanding of the basic procedures and a benchmark against which the audit team can make comparisons to determine compliance status after the team has gained an accurate understanding of the facility's management approach and pro-

FIGURE 3-1. Basic steps in the typical audit process.

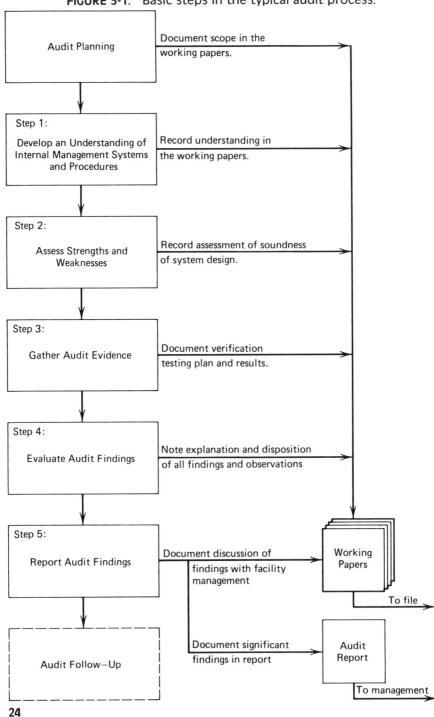

cesses. Similarly, many companies have developed fairly extensive written hazardous waste management programs that outline responsibilities and accountabilities, inspection procedures, training programs, and other aspects of the organization's internal control system for hazardous waste management. Such plans can be especially useful to the auditor in understanding the facility's intended management approach.

2. *Assessing Strengths and Weaknesses.* The second step in the on-site audit process involves assessing the strengths and weaknesses of those internal management procedures and systems identified and described in Step 1. Here, the auditor looks for indicators such as clearly defined responsibilities, an adequate system of authorizations, the awareness and capability of personnel, documentation and record keeping, and internal verification. This step provides the rationale for conducting subsequent audit steps. In situations where the design of the internal environmental management system is assessed as sound (that is, acceptable results will be achieved if the system is functioning as designed), the subsequent steps of the audit can focus on the effectiveness with which that design is implemented, and the extent to which the system in fact performs as intended. If the design of the internal system is not sound enough to ensure the desired results, the subsequent audit steps must focus on the environmental results (e.g., compliance status, safety practices and records, etc.) rather than on the internal management system. In other words, it is unacceptable for environmental auditors to focus their examination on the functioning of an internal management system which they have assessed to be flawed in design.

3. *Gathering Audit Evidence.* The third step in the audit, gathering audit evidence, serves as the basis upon which the audit team determines compliance and forms its audit opinion. Suspected weaknesses in the management system are confirmed in this step; systems which appear to be sound are tested to verify that they work as planned and are used consistently. Audit evidence can be collected through inquiry (formal questionnaires, informal discussions), observation (physical examination), and testing (retracing data, verifying paper trails, etc.). Many companies follow formal audit checklists and guidelines, and some companies actually conduct effluent sampling and analysis. The audit team should identify and then verify those activities in the environmental management process that can give the greatest insights into the overall functioning of the system. Audit evidence (as described in Chapter 11) can be physical, documentary, or circumstantial.

4. *Evaluating Audit Findings.* After audit evidence has been gathered, audit observations and findings are evaluated. The purpose of this

step is to assimilate and integrate the findings and observations of each team member and then determine their ultimate disposition—either included in the formal audit report or only brought to the attention of facility management. Typically this is done through meetings of the audit team members prior to the completion of the audit. Here findings and observations can be organized to determine whether there are common findings that, when viewed as a group, may have greater significance than they do individually. In evaluating audit findings, the team members, and particularly the team leader, determine whether the audit evidence is sufficient to support audit findings and whether some or all of the supporting examples should be included in the report.

5. *Reporting Audit Findings.* The environmental audit reporting process often begins with informal discussions between the auditors and the facility environmental coordinator when discrepancies are noted. Findings are further clarified as the audit continues and then reported to facility management during the audit exit meeting or close-out conference. During the meeting, the audit team communicates all findings and observations noted during the audit and indicates which items will appear in the formal audit report.

The purpose and uses of the audit report include providing management information, initiating corrective action, and providing documentation of the audit. Most companies have a formal written report of the audit prepared by the audit team leader with input from the team members. The audit report provides appropriate linkages for all review learnings so that the existing management framework is able to determine what, if any, actions are required. Some companies use a multiple or hierarchical reporting scheme. Whichever approach is used, an effective reporting process will communicate the audit results and identify issues to the appropriate persons within the company.

POST-AUDIT ACTIVITIES

The audit process does not end at the conclusion of the on-site audit. Typically, the audit team leader prepares a draft report of the findings and observations within two weeks of the on-site audit. This draft report may be reviewed by the Environmental Affairs Department, the Legal Department, facility management, and so on, before a final report is issued. As the final report is being prepared, the action planning process is usually initiated. This process includes determining potential solutions and preparing recommendations, assigning responsibility for corrective action, and establishing timetables. The final step in the overall audit process commences with the follow-up to the action plan to ensure that all deficiencies have, in fact, been corrected.

PART TWO

DESIGNING AN AUDIT PROGRAM FOR YOUR COMPANY

CHAPTER 4

SELECTING AUDIT GOALS RESPONSIVE TO YOUR ORGANIZATION

Selecting and defining explicit goals and objectives for an environmental auditing program is seldom easy. Ultimately, audit program goals are best selected when the full spectrum of company and management needs is considered. For the convenience of the reader, Chapter 4 briefly summarizes the more common audit goals, described in Chapter 2, before discussing several key issues to consider in selecting from among those goals.

BASIC AUDIT GOALS

Auditing serves needs beyond those of evaluating and verifying compliance. The objectives vary widely, depending on the company's culture, management philosophy, and size. (See Table 4-1.) Program objectives commonly found include providing assurance to top management, helping facility management understand requirements and identify compliance problems, identifying ways to optimize resources, and identi-

TABLE 4-1. POSSIBLE AUDIT PROGRAM GOALS AND OBJECTIVES

Identification and documentation of compliance status, including
compliance discrepancies,
differences or shortcomings at individual facilities, and
patterns of deficiencies that may emerge over time.

Improvement in overall environmental performance at the operating facilities
as a result of
providing an incentive not to allow problems to happen again,
reducing or containing problems that can interfere with normal business
activity, and
providing an incentive to clean up or improve housekeeping before an upcoming audit.

Assistance to facility management in
understanding and interpreting regulatory requirements, company policies
and guidelines, and (perhaps) good practices,
identifying compliance problems,
defining cost-effective measures that should be taken to achieve compliance, and
putting potential problems before a "committee of experts."

An increase in the overall level of environmental awareness as a result of
demonstrating top management commitment to environmental compliance,
increasing the environmental awareness at the facility,
the training accrued to the audit team, and
involving employees in environmental, health, and safety issues.

Acceleration of the overall development of environmental management and
control systems as a result of
auditing those systems that are "auditable,"
defining the status of those activities that are not yet in a position to be
audited,
identifying important lessons learned and modifying systems and/or sharing
information as appropriate,
encouraging formulation of more formal procedures and standards for measuring environmental performance, and
developing a data base of information on environmental performance that can
be used in other management functions.

Improvement of the environmental risk management system as a result of
identifying conditions that may have an adverse impact on the company,
assessing the risks associated with the hazardous conditions identified, and
determining what actions are necessary to control those risks.

Corporate protection from potential liabilities as a result of
being able to demonstrate due diligence or evidence of the corporate environmental commitment,

TABLE 4-1. *(Continued)*

soliciting an independent (third-party) opinion,
documenting evidence demonstrating that the company is complying with
regulations, and
developing improved relations with regulators.
Development of a basis for optimizing resources as a result of
identifying current and anticipated costs and recommending ways to reduce
those costs,
identifying potential longer-term savings that can be accrued, and
identifying potential opportunities to reduce waste generation.

---

fying hazardous conditions that may have an adverse impact on the
company.

While an audit program may have several basic purposes—all individually worthwhile—no single company is likely to be able to devote the
resources to an environmental audit program that would be required to
fulfill all or even most of them. Because these program objectives are
often competing and conflicting in terms of the role of the audit team, the
audit methodology, and the type of audit reporting required, it is important to select audit goals that are responsive to your organization.

KEY ISSUES TO CONSIDER

In selecting audit program goals, one must pay careful attention to three
key issues: the needs of top management, the corporate philosophy and
culture, and the distinction between short-term and long-term goals.
Each of these issues is important whether the motivation for the program
comes from upper, middle, or lower management.

Top Management Needs

In general, the desires and needs of the board of directors and the CEO
greatly influence both the thrust and character of the environmental
management system and the specific purpose and role of environmental
auditing within that system. Many corporate environmental auditing
programs are initiated at the request or prompting of top management.
Rigorous, sophisticated audit programs often exist because the board of
directors and/or top management believe corporate environmental pro-

tection is important and necessary. And even in those companies where the driving force behind the audit program is middle management, it is critical to manage the program in a manner consistent with the needs of top management.

The needs (and responsibilities) of top management and the board of directors have changed significantly. For example, boards of directors initially represented the owners of the company. As corporations grew and went public, the typical board became staffed with members of the company's management. Its role became what the president of the company wanted it to be. In recent years, as a result of court rulings on director liability and the expanded scope of the director's responsibility, outside directors have become a prominent force on many boards. Board members now want assurances that they are directing an organization that is a "good corporate citizen" and is also controlling costs to protect the stockholders' interest. And, they want to *know* the corporation's compliance status. Auditing can help assure the board that compliance is being achieved and there are no material risks.

Top management must manage the corporation within the prescribed policy of the board. Moreover, the CEO is charged with managing the organization efficiently, profitably, and responsibly. The environmental management system supports these needs and an audit program provides feedback and serves as a check on that system. In essence, an effective auditing program that is integrated throughout all levels of the organization can help top management know that its mandate is being followed— that operations are consistent with policy. A voluntary environmental audit program can also help provide evidence of due diligence, that is, that management is exercising appropriate care and attention in discharging its duties.

The needs of the environmental manager, facility management, and audit team members also should be identified and considered. In many companies, environmental auditing is seen by environmental and operating management as a powerful management tool. To the environmental manager, an audit program is a way of gaining assurance that operations are in compliance. To operating management, an environmental audit is often seen as having an "outside set of eyes" to look over the environmental aspects of a facility.

Corporate Philosophy and Culture

It is critical that the specific goals of your audit program be consistent with the culture, values, norms, and overall environmental management philosophy of your company.

In sorting out and defining the corporate philosophy, it is useful to recall the three-stage model of environmental management described in Chapter 1. In Stage I, where the principal thrust of environmental management is to stay out of trouble, the "audit program" (if any) is likely to be mostly an assessment (or judgment) of the facility's potential environmental problems. Very few specific standards and procedures are likely to have been developed either for monitoring and recording environmental data or for taking extra steps to assure that operations are consistently conducted in an environmentally acceptable manner.

In Stage II, the basic focus shifts to managing compliance and the company begins to develop a system of policies and procedures to help ensure that desired compliance levels are being achieved. An environmental audit program may be established to provide feedback on how compliance is being managed. Moreover, with more standards (as well as regulatory requirements and company policies) to audit against, those conducting the audit can begin to test or verify the compliance status.

In Stage III, the thrust is one of risk management and the focus of auditing typically shifts to include an audit of the management system, usually in recognition that the company is vulnerable not only to those risks associated with noncompliance, but also to environmental risks not (at least yet) covered by specific compliance requirements. Here, the audit program may also have an added objective of seeking to identify gaps in the management system where unanticipated hazards could have "material" impact on the corporation.

Whatever the corporate philosophy, the audit program objectives should be consistent with the overall philosophy of the corporation.

Distinction Between Short-Term and Long-Term Goals

The purpose of an environmental audit program should be considered from the perspective of its long-term goals, short-term goals, and specific objectives.

Long-Term Goals.   What do you envision as the overall roles of the environmental audit program looking out over the longer term? (These include those ends that you do not expect to attain within the current planning horizon, but hope to attain later and toward which you expect to make early progress.) For example, you may establish a goal of confirming that systems are in place and functioning to identify and manage all environmental hazards. Or you may desire to work toward developing an overall environmental performance tracking system that would be periodically audited. Similarly, you may wish to work toward developing and attaining your own internal standards that would be confirmed by a periodic audit. Whatever your particular longer-term auditing and environmental management goals are, it is important to pay attention to the particular skills and capabilities that will need to be developed.

Short-Term Goals.   What goals need to be attained during the current planning horizon (especially one year)? For example, you may identify as a short-term goal the confirmation that *all* air emission sources are properly registered as required by applicable federal, state, and local requirements. Such a short-term goal might require focusing audit resources (perhaps at the expense of other audit areas) on directly confirming the registration or permit coverage of every air source.

Specific Objectives.   What specific measurable objectives must be met to fulfill your stated short-term goals? These can include both program objectives (e.g., how many audits, what functional areas, what resources) and individual audit objectives.

THE SELECTION PROCESS

In sorting through the potential goals for your environmental audit program, carefully analyze the implications of each of the potential goals in terms of the role of the audit team, the audit methodology, and the type of audit reporting required. Remember that an audit program can have different purposes, all legitimate, but no individual company is likely to be able to devote the resources required to fulfill all, or even most, of them.

In selecting audit program goals responsive to your organization, ensure that the goals selected are as follows:

Meaningful.   Do the goals and objectives defined fill an important role in the overall environmental management objectives of the company?

Operational.   Are the objectives you set reasonable? Are the chances better than average that you will be able to attain the goals and objectives within the time frame established?

Specific and Measurable.   Are the objectives specific enough to be measurable?

Controllable.   Is the fulfillment of the goals within your control?

Directed.   Are the goals and objectives directed toward a specific action or end result?

After the necessary trade-offs have been made and the audit program objectives have been defined, a policy statement defining the program's purpose, responsibilities, and authority can help minimize future ambiguities that may interfere with achievement of audit goals and objectives.

It is important to recognize that selecting audit program goals and objectives responsive to your organization is a dynamic process. Just as the stated objectives influence the design and implementation of audit programs, the experiences gained in implementation can help reshape those objectives. For example, one unanticipated result of an audit program might be that employees are better trained in understanding the regulatory and corporate environmental requirements. While training may not have been an objective of the audit program initially, it may gain importance and ultimately become one of the objectives of the program.

Chapters 5 through 8 discuss a number of important issues involved in designing an audit program for your company. As you read those chapters, continually work toward defining in more concrete terms the environmental audit program goals that are most responsive to your organization. Part 3 discusses specific techniques to consider including in your program. Selection among the various approaches will be easier if you develop explicit goals and objectives.

CHAPTER 5

DEFINING THE BOUNDARIES OF YOUR ENVIRONMENTAL AUDIT PROGRAM

One of the greatest challenges in designing and developing an environmental audit program is to convert the general program goals into specific program boundaries that both support those goals and are realistic. For any given set of objectives, the audit program can take a variety of shapes, depending upon how the individual company balances the options available.

In choosing among options, it is important to distinguish between the boundaries of an environmental audit program and those of an individual audit. While a broad array of activities may be included in an audit program, the focus of a single audit may be much narrower. For example, an audit program may cover air pollution control, water pollution control, and solid and hazardous waste management, whereas the scope of an individual audit may be limited to hazardous waste. As another example, the audit program may include the full range of environmental, health, and safety issues while the scope of an individual audit could be limited to only one or two disciplines.

In defining the boundaries of an audit program (and of an individual audit), the program manager should seek to answer three questions.

What should the scope and focus of an environmental audit program include?

When is an individual audit complete?

How can boundaries be selected to optimize the resources and enhance the effectiveness of the overall audit program and individual facility audits?

The specific answers to these three questions will vary with the needs, objectives, and available resources of each individual company.

WHAT SHOULD THE SCOPE AND FOCUS INCLUDE?

The scope and focus of an environmental audit program can be defined in the context of four separate parameters: organizational, functional, compliance hierarchy, and locational boundaries. Options to consider for each are discussed below.

Organizational

Determining the organizational boundaries is generally straightforward. Here the program manager must identify which parts of the organization the audit program is to serve. Is the program sponsored, organized, and conducted for the benefit of the entire corporation or for a particular division or operating unit? If a corporate-wide program, should it include all divisions (or only particular divisions); all operating units (or only specific operations such as chemical plants, laboratories, terminals, etc.); all activities (or just those that require specific environmental attention)?

The selection of organizational boundaries is governed by the purpose of the audit program, the needs of the individual(s) who have caused the establishment of the audit program, the culture of the corporation, the relationships between the individual divisions and operations, and the structure and reporting relationships of the environmental organization.

Functional

A number of specific environmental disciplines or functional areas can be included in the scope of an environmental audit program. Usually included are air pollution control, water pollution control, and solid and hazardous waste management. Some companies also include other areas such as occupational safety, occupational health, product safety, medical, and transportation. A few companies have defined their program scope to include all regulated activities and, therefore, also include such items as equal employment opportunity requirements. Depending on the nature of the corporation's operations, such activities as Food and Drug Administration requirements, Bureau of Alcohol, Tobacco, and Firearms issues, and the like may also be included.

In determining the functional boundaries it is important to be as specific as possible about the subject areas to be included. For example, water pollution control can be subdivided into such areas as permits, discharges, sampling, spill prevention and control, groundwater monitoring, and drinking water. Water sampling procedures can be further broken down into sampling technique, analysis, documentation, calculation of results, preservation of samples, and so on. An audit can include all of these or just one. Some audit programs explicitly exclude the general safety and health considerations but include employee occupational health exposures and hazards related to chemical substances. Some audit programs include employee noise exposures in the workplace but exclude other Occupational Safety and Health Administration (OSHA) considerations.

Compliance Hierarchy

Once the functional areas are defined, the specific compliance parameters must be determined. The audit may be directed toward establishing compliance with federal laws, regulations, and standards; state regulations; and/or local regulations. Many companies also include compliance with policies, procedures, and guidelines issued at various levels within the organization (corporate, division, or facility). Occasionally, they include other compliance parameters, such as industry standards.

Locational

Selecting the locational boundaries of an environmental audit at a specific site can be very complicated. Certainly, those boundaries must include all activities within the facility's property line. However, off-site activities may need to be included as well, and these may be difficult to define. For example, how far along the transportation routes to off-site hazardous waste disposal facilities should an environmental audit extend? Which off-site areas should be examined? Should the audit include the impact of the facility on nearby environmentally sensitive areas, such as a lake or stream, or other sensitive areas such as residential areas, schools, or recreation facilities? Should ambient air monitoring off-site be included in the audit? Should the audit also include toll manufacturing or contract packaging operations conducted off-site in support of the facility being audited?

Few facilities function in isolation from other units within the organization or from other organizational entities. The challenge in establishing an audit program is to define the locational boundaries in a way that fulfills the audit program objectives without spreading limited resources too thinly. One guideline is to recognize that there are two ways to audit activities beyond the actual facility property line. One is for the auditor to physically go off-site. The second is to review information available on-site.

WHEN IS THE ENVIRONMENTAL AUDIT COMPLETE?

The boundaries of an audit must be defined clearly and appropriately so that the audit program does not become overwhelming or impossible to manage. Audit breadth must be traded off against audit depth.

Audit breadth refers to the range of activities included in the on-site audit and relates to the functional, compliance, and locational factors described above. What specific areas are to be covered in an audit of a given facility? Will the scope of the individual audit include all of the functional areas defined for the audit program or just selected ones? Is the team expected to look at just those activities within the facility boundaries or a number of others outside the fence as well?

Audit depth refers to the level of detail. The critical decision is to determine which audit steps or tasks need to be examined in what level of detail so that the recipient of the audit report, the audit team leader, and each of the team members are satisfied that the audit supports the goal of the audit program.

Some balance must be achieved between audit breadth and depth. An in-depth examination of many specific areas to assure a high probability that *all* operations are in compliance with *all* applicable regulations, policies, and procedures probably requires an inordinate effort. A more common practice is to choose a representative sample of activities to be audited, identify the specific period to be reviewed, determine the compliance status of the sample, and focus the audit on establishing whether appropriate management and control systems are in place to assure that compliance is consistently achieved over the period chosen. This approach permits the auditors to examine a sample of the facility's activities and records in detail.

IS RESOURCE USE OPTIMUM?

For a fixed allocation of resources there is some set of boundaries, in terms of scope, focus, breadth, and depth, that will ultimately lead to an audit program of maximum effectiveness. Alternatively, given a set of audit program goals and cost-effectiveness targets, some resource allocation will give optimum results.

Generally speaking, defining program boundaries is an iterative process whereby boundaries are selected on a trial basis and modified as experience is gained. Several aspects of this iterative process are particularly noteworthy.

Start Small

Do not spread the audit program too thinly. Trying to include too many facilities or too many compliance parameters with a limited amount of resources is sure to inhibit the effectiveness of the audit. It is always easier to expand the effort on the second round than to be frustrated on the initial audit.

Try Different Sets of Boundaries

It may be advisable, both in starting an audit program and occasionally after the program is up and running, to select different boundary conditions for different audits. For example, one audit initially may be limited to on-site activities; another may include off-site operations as well. For a PCB audit, the team can look at on-site activities (inventory, storage, marking, labeling, etc.), or can include examination of outside laboratory or waste disposal facilities. Using different sets of boundaries will help determine the approach that is likely to be most effective for your organization or for a given facility.

Carefully Consider the Trade-Offs in Combining Functions

There are both advantages and disadvantages to combining several functional areas in a single audit. Combining areas may economize audit team

TABLE 5-1. PROFILE OF AUDIT PROGRAM BOUNDARIES

	Company A	Company B	Company C
Primary program goal	To improve environmental performance, reduce costs, and verify compliance	To help facility managers achieve environmental, health, and safety compliance	Assess the internal management and control system; provide a vehicle for training managers
Audit program parameters			
Organizational	Major manufacturing locations only	All manufacturing locations, warehouses, and distribution centers	All manufacturing locations
Functional	Solid and hazardous waste and PCBs	Air, water, solid and hazardous waste, spill prevention control and countermeasures, occupational safety, occupational health, product safety, and transportation	Air, water, solid and hazardous waste, and occupational safety and health
Compliance	Federal, state, and local regulations; corporate policies and procedures	Federal, state, and local regulations; company and plant policies and procedures	Federal, state, and local regulations; corporate, division, and facility policies and procedures
Locational	On-site activities only	On-site activities; off-site disposal sites	On-site activities; selected off-site activities
Individual audit parameters	All audits include the program parameters defined above	All audits include the program parameters defined above	Although the parameters of the audit are broad, each audit scope is site specific. For example, one facility audit might include only air pollution control while another facility audit might focus only on occupational health

resources, use available facility resources more effectively, and provide for sharing of audit functions among the audit team members. On the other hand, including too many functional areas in a single audit may require too large a team or too many facility resources for effective operation. This can be an especially important consideration where the facility has a particularly small staff. A broad scope or large audit team can "overpower" the facility staff. In such situations the audit team can find itself facing a bottleneck where the team members literally stand in line waiting to talk to the facility's environmental coordinator.

Be Prepared

The audit team must do its homework. A basic understanding of the facility, its organization, processes, and operations can assure a more effective relationship between the audit team and the key staff of the facility being audited as well as help conserve resources. Similarly, familiarity with company policies and regulatory requirements prior to the audit is essential in maximizing the effectiveness of the team. Chapter 9 discusses pre-audit activities and audit planning in more detail.

SELECTING APPROPRIATE BOUNDARIES

No single prescribed set of boundaries can guarantee an effective audit program. Boundaries must be developed in concert with the goals and objectives discussed in Chapter 4, the resources available, the nature of the organization, and the environmental hazard potential associated with different parts of the operations.

Different companies have adopted very different sets of program boundaries. Table 5-1 outlines those used by three companies.

Selecting audit program boundaries appropriate for a given company often takes at least two to three years of experimentation with different audit approaches. And, even when those "optimum" boundaries are determined, they will demand subsequent review and evaluation to make sure the specific scope, focus, breadth, and depth of the individual audits consistently achieve the goals set for the program.

CHAPTER 6

ORGANIZATION AND STAFFING ISSUES

The appropriate organization and staffing of an environmental audit program revolve around the need to balance objectivity and familiarity with the operations undergoing audit and the requirements against which they are audited, the need for competence in environmental auditing skills and techniques, and the need for credibility with both the audited operations and senior management.

These needs suggest three basic questions that should be answered in choosing among the organizational and staffing alternatives for your environmental auditing program.

Organizational Location. Should the environmental audit program be housed within the corporate environmental, health, and safety affairs department or elsewhere within our company?

Program Management and Staffing. What basic program management and staffing considerations need to be provided for?

Audit Team Characteristics. Of the wide range of expertise available, which team member characteristics are essential, which are desirable, and which are not desired?

In addition, the management of the overall quality of the audit program is closely linked to the program organization and staffing. That is, the organizational unit chosen to house the audit program and the manner in which your audit teams are staffed will provide the framework and basic context in which audit quality is measured and managed. Therefore, an additional question should be considered in choosing among the organizational and staffing alternatives.

Managing Program Quality. What is the best way to organize and staff the audit program so that it provides a sound framework for managing audit quality?

Chapter 6 presents the basic organizational and staffing alternatives. Chapter 20 presents basic aspects of managing audit quality.

ORGANIZATIONAL LOCATION

The question of where the audit program should be housed is basically one of the "distance" needed or desired between the management of environmental, health, and safety affairs and the program that will review or audit these management processes. Some companies seek independence of the audit function while others place a premium on access to and familiarity with information that is to be audited. Other considerations include the availability and source of the resources needed to carry out the program, the nature of ongoing linkages needed with other parts of the organization, and the need for continuing development, refinement, and management of the program.

Most environmental audit programs are housed within the corporate environmental, health, and safety affairs department. In fact, we estimate that more than 90% of the companies with environmental audit programs have placed their program within the corporate environmental affairs department. However, the location of the environmental audit program within corporate environmental affairs should not be viewed as a foregone conclusion. Other organizational locations include the company's internal audit department, regulatory affairs department, production or operations, and the law department. Companies that have located their programs in some part of the organization other than environmental affairs have done so for some important reasons.

Companies that select the internal audit department to house the environmental audit program tend to view the program more as a corporate (rather than environmental) management tool. In these companies, the audit program serves strictly as a double check on the system. Operating management is expected to manage environmental hazards and compliance while environmental management is expected to provide guidance, direction, and oversight. The environmental audit is conducted by individuals who are not part of the day-to-day environmental management and provides an independent spot-check on the system. Staffing approaches vary among those programs housed in the internal audit department. In a few cases, the staffing emphasis is on auditing (and not environmental) and the work is conducted largely by individuals with general training in operational auditing. In other cases, the emphasis is equally on environmental and auditing with a technically trained staff.

Production or operations departments are often chosen as the location of the environmental audit program to reflect and reinforce the company's philosophy that operations is responsible for environmental management. The legal department is selected because of the sensitive legal issues involved.

PROGRAM MANAGEMENT AND STAFFING

Wherever the environmental audit program is housed within your organization, there are two important considerations relating to program management and staffing. First, what are the basic program management responsibilities? Second, what is the proper level of commitment and availability of the audit staff?

Program Responsibilities

Whether or not there is a full-time audit program manager, there are a number of basic program management responsibilities that include the following:

Resource Management.   Managing the resources available to the program, including the budget, the personnel, and the other resources available from other parts of the organization.

Staff Selection and Training.   Selecting, orienting, training, and continually developing the audit team staff.

Program Development.   Continuing to develop, refine, and advance the audit program.

Keeping Current.   Keeping up to date on current activities in the field.

Most audit programs interact with, and even depend on, other parts of the organization. For example, often a variety of organizational entities are drawn on for: audit team staffing, program resources, specialized expertise, corrective action, and reporting on program status. Thus, regardless of where the audit program is housed, it will not operate in a vacuum, and one of the primary responsibilities of the program manager is to provide and utilize those organizational linkages.

Commitment and Availability of Audit Staff

The question of full-time or part-time program management, and of full-time versus rotating staff, is largely one of continuity and focus, balanced against the scope and focus of the audit program and the resources required to fulfill the program goals and objectives. Here there are three basic alternatives.

A full-time audit manager and full-time audit staff.

A full-time audit manager with staff assigned on a collateral basis.

Both audit manager and staff assigned on a collateral basis.

In selecting from among these alternatives, one must consider a number of factors. The principal advantage of full-time personnel is continuity from audit to audit. While it is highly desirable, and often essential, to have a full-time audit program manager, it is possible to have a very sound program with either full-time or part-time staff within the auditing group. If the staff are to be part-time, it needs to be explicitly understood whether those individuals will be on a continuing part-time basis with the auditing program (where the same staff are used on a periodic basis) or on a rotating membership basis (where new staff are continually used). A permanent, continuing basis makes the conduct of facility audits much more effective, and ultimately more efficient, and is more likely to lead to an audit approach that inspires greater confidence that the goals and objectives of the program are being met. Rotating membership can have very positive benefits in exposing, and thereby training, a wide range of people to the auditing program. But it does require continually training auditors and searching for people who will bring to the program the dedication, competence, and interest that is required.

AUDIT TEAM CHARACTERISTICS

As is the case with other types of auditing, the quality of an environmental audit is only as good as the quality of the individuals conducting the audit. The challenge for the audit program manager is to select, orient, train, and manage the audit team members in a way that ensures the audit program objectives will be met.

One of the dilemmas the audit program manager faces is that different people within the organization have very different views on who the "right" people are to conduct an audit. If the program manager were to ask facility management who should *not* conduct an environmental audit of an operating facility, the four most frequent answers would probably be outside consultants, corporate auditors, corporate lawyers, and corporate environmental managers. Most probably, the most desired auditor would be the facility manager.

Now, consider the board of directors who have responsibilities to stockholders and who want to have some independent view of how the corporation is performing in the environmental area. If one were to ask the board of directors who provides the most credible independent view of a facility's performance, most likely the four most frequent answers would be outside consultants, corporate auditors, corporate lawyers, and corporate environmental mangers. And, who is the least credible to the board with regard to environmental management and environmental auditing? The facility manager is.

Herein lies the dilemma. The most credible auditors to the board are the least desired by the facility, and the most desired by the facility are the least credible to the board. Given this dilemma, the audit program manager needs to be sure that the staffing objectives are supportive of, and consistent with, the overall goals of the audit program.

In sorting through the many staffing options available, it is useful to begin by identifying the desired areas of expertise of the audit team, and then look at the specific team leader and team member characteristics.

Desired Areas of Expertise

Legal Considerations.   First, and most important, the audit team members need to know the laws, regulations, and permits that apply to the facility being audited. It is essential to know the federal, state, and local regulatory standards you are auditing against. In addition, it is essential to know what corporate policies and procedures apply if you are auditing against them.

Relevant Environmental Control Technologies.   In like manner, understanding the environmental, health, and safety control technologies that fit the facility's operations and the kinds of process effluents and waste that the facility generates is an important component of the expertise of the team. Having expertise in the functional area being audited is important.

Scientific Disciplines.   To the extent that the auditing team is expected to identify hazards that have not been identified previously or have been considered to be unimportant, it is crucial to have on the team technical experts who would recognize the potential for hazard in a particular situation or operation. This could range from the potential for groundwater contamination, to impacting nearby foliage, to the presence of a particular chemical substance that, because of past practice, has been considered to be unimportant but, given today's knowledge, could have major significance.

Facility Operations.   Another important area of expertise is an understanding of the facility operations and processes. Even if the overall process is the same among selected facilities, each facility has some differences. It is generally possible to search for and use people within the company who have some familiarity with the kinds of operations that are going on in the facility being reviewed. It is especially helpful to include on the team one or more people who understand the type of facility being audited. Knowledge of facility operations makes it easier for the auditor to relate to facility personnel and provides a basis for better understanding what facility personnel say and show the auditor.

Auditing.   As environmental audit programs develop, and particularly when one of the purposes of the program is to provide assurance to management, expertise in the discipline of auditing becomes increasingly important. This includes training in verification techniques, audit procedures, and other aspects of the internal auditing disciplines and a general awareness of the techniques and practices used by other companies.

Management Systems.   An understanding of management systems—both environmental management systems and also facility management systems—can be especially important as an auditing program places increased emphasis on verification. Management systems include information on how work is planned, implemented, controlled, and reviewed; the roles and responsibilities of various managers and functional disciplines; and reporting relationships and how information flows both within the facility and from the facility upward in the organization. Gaining an understanding of how the facility manages itself and how it is managed by the corporation has an impact on the effectiveness with which the team determines the extent to which the management systems are in place and working.

Knowledge of Peer Facilities.   Finally, having an understanding of what similar companies and facilities are doing to control hazards and manage for compliance can be very helpful to the audit team.

Team Member Characteristics

The choice of team members is important. Look at the following characteristics:

1. *Experience*.   The composition of the team is very directly related to the function that the team member is to perform during the audit. If it is desired that the team assess environmental performance and identify hazards, then technically trained people (engineers, chemists, safety professionals, industrial hygienists, and so on) who have that experience and expertise are desirable team members. If the focus of the audit is to verify records and find that procedures and policies are being followed, then people with experience in auditing, and not necessarily as much experience in environmental control techniques, are good team members.

2. *Training*.   Have team members participated in, or been exposed to, an actual environmental audit, especially at a facility similar to the one to be audited? Training or familiarity with not only the company's audit objectives but also the company's established auditing practices and approaches can be very important. Either training should be conducted prior to the audit, or extra time and resources should be provided during the audit to train new members of the audit team. Team members should be prepared to conduct an audit efficiently and effectively while at the facility. This is especially true if the team members are on a rotating basis. As crucial as pre-audit training is for the auditors, actual on-the-job experience can be the best teacher, as with so many other kinds of situations. Therefore, after team members have done some audits, they will become better auditors and more effective at the facility.

3. *Independence*.   The degree of audit independence that is desired by the person for whom the audit is being conducted will influence who is on the team. If an inside, independent verification is sufficient, no outsiders need to be involved. This has been the view of many of the programs currently being conducted. However, if the board, corporate management, or environmental management decides that they would like to have an outsider's view of the compliance status of the corporation's facilities, it is necessary to go outside the corporation, for no one within the corporation can provide the independence sought. (Note: Some companies look to outsiders primarily because they lack available in-house expertise rather than for added independence.)

4. *Sensitivity*.   While an understanding of the regulatory requirements is of utmost importance, it is equally crucial for audit team members to be sensitive to the issues and concerns about identifying potential violations of laws and regulations or findings about inherent potential hazards that have not yet been controlled or for which there are no plans for control. An auditor should not set up a "smoking gun" situation where a regulatory exception is identified but not followed up properly. In addition, each individual auditor has an obligation to the auditing program to not duck any tough issues. Moreover, it is as much a disservice to the corporation to state that the facility is in violation of regulatory requirements when in fact no law or regulation yet exists as it is to state that the facility is only in violation of corporate policy when, in fact, it is in violation of a regulation.

Team Leader Characteristics

In addition to the team member characteristics described above, questions arise about the special characteristics the team leader should have and what kind of experience he or she should bring to bear on the audit program. Typically, team leaders are technically trained personnel, engineers most frequently, who have had experience in environmental management and extensive experience in at least one functional area, and possibly more. Because of experience, most, if not all, team leaders have a general familiarity with facility operations.

The team leader needs to have the support and respect of management, the audit team, and facility personnel. If the audit program is designed to serve the needs of top management, those individuals need to be assured that the team leader will conduct a comprehensive audit. To maximize the effectiveness of the on-site review, the team leader needs to be someone who can command the respect of the other team members. In addition, facility personnel, quite understandably, may be very concerned about being audited and what the audit team is going to do in the facility and what they are going to say about the facility to their bosses. In short, the team leader should be believable to the facility management, should have the capability and willingness to help the facility understand why the facility is being audited, what the audit is all about, and what the audit team is doing. And, the team leader must recognize the fine balance between not interfering with the facility's primary mission, of producing a product, and not being so supportive that the team does not, in fact, do the rigorous audit that generally needs to be done.

CHAPTER 7

DESIGNING A REPORTING PROCESS

Recording audit findings and reporting them to the appropriate level of management are key to any effective environmental audit program. Chapter 7 first discusses some perspectives and options to consider in developing a reporting system and then presents key issues to consider in designing a specific reporting system.

PERSPECTIVES TO CONSIDER IN DESIGNING A REPORTING SYSTEM

Designing an effective environmental audit reporting process requires taking into account and balancing the differing and sometimes conflicting needs of various perspectives within the corporation. Each of the needs described below is legitimate, and none should be overemphasized at the expense of others.

Board of Directors

Board members have the special obligation to inquire into and receive information on the environmental performance of the organization. In

general, their first concern is with the adequacy with which environmental problems are being managed. Thus, they want answers to questions such as, Does our environmental management program compare
favorably to other companies' programs? Are any environmental matters
likely to have a material impact on the corporation? Are we in compliance
with federal, state, and local environmental laws and regulations?

If the answers to such questions are less favorable than they would like,
they need to hear that management is actively addressing environmental
problems with a specific timetable for resolving them in a way that
benefits the corporation financially and in terms of public image. The
board wants assurances that they are directing an organization that is a
"good corporate citizen." But, in a possible conflict of desires, they also
want to hear that costs are being controlled to protect the stockholders'
interests.

In short, board members are less concerned with the details of environmental matters than with the potential effects on profits and losses to the
corporation, with the likelihood of significant harm to the environment,
and with the corporation's and their own personal legal liabilities.

Corporate Officers

Corporate officers need to be assured that the environmental management functions are being conducted in accordance with applicable laws
and regulations. They need to know that managers have taken responsible charge of environmental matters and are fulfilling the corporate
obligations. Corporate officers generally want to hear "good news" in
environmental areas, to have an awareness of areas of "bad news," and to
know that problems are being attended to at appropriate managerial
levels. They do not want to hear about "Band-Aid" measures for life-
threatening problems.

Corporate officers need more detail than board members since they
have line responsibility for the organization. Yet, their desire for detail is
often to provide greater assurance that they will not be caught uninformed or without adequate knowledge of real problems and the remedial
actions undertaken to alleviate them.

Operating Management

Operating management, with overall responsibility for all aspects of
assigned operations, needs to know whether any serious environmental
problems need attention. They want to know about any situations that

could curtail or otherwise negatively impact operations. Like corporate management, operating management generally is not interested in all the details of day-to-day environmental compliance. They generally do not want to hear about extensive environmental "fixes" that will require large amounts of capital and/or operating costs unless the equipment can be tied to either a specific requirement or a long-term operating benefit.

They do want to know what is going to be reported to corporate management *before* it is reported (and usually want slightly greater detail than will be reported to corporate management so that they are in a position to respond to any questions). Operating management also wants assurances that environmental matters within their areas of responsibility are receiving appropriate attention. Where the development of environmental management systems is largely left up to operating management rather than somewhat standardized across the corporation, operating management may also look for confirmation that their environmental systems and procedures are appropriate. Depending on what environmental staff resources and capabilities are available within their division and at their facilities, they may also want to know what corrective actions are believed most appropriate by the auditors and may look for assistance in correcting deficiencies noted during the audit.

Environmental Management

Environmental management's basic need is to understand the exceptions or problems noted during the audit and the specific observations of deficiencies or discrepancies on which they are based, so that they can correct the problem and not just the symptom. This understanding is important not only for their particular role in the problem-solving process but also so that "lessons learned" from the audits can be communicated elsewhere within the company.

Other needs of environmental management will vary considerably from company to company depending on the corporation's basic environmental management philosophy, the role of the environmental affairs staff, and the relationship between environmental management and the audit program. In some companies, for example, the programs can be perceived as an audit of the performance of the corporate or division environmental affairs staff as well as an audit of the facility's environmental performance.

One or both of two factors can lead to this situation (1) a prominent role by the corporate or division environmental affairs staff in the day-to-day environmental management of the facility or (2) location of the audit program outside the environmental affairs staff in question—either else-where in the company or in the case of a division environmental staff, within the corporate environmental affairs organization. In such situ-ations, environmental management needs to be in a position to disagree with or take exception to the findings of the audit.

In other companies, line management alone is held fully accountable for the environmental performance. Environmental management is viewed as a full partner (if not primary owner) of the audit program. Here, environmental management is less apt to be defensive about the excep-tions found during the audit and will want to know if there are any unrecognized weaknesses or "blind spots" in the company's environmen-tal management program. They are looking for confirmation that the systems are in place and functioning to manage appropriately the environ-mental aspects of the corporation's operations.

Facility Management

Facility management (and facility environmental staff), as "auditees," want and *expect* fair and equitable treatment in the environmental audit reporting process. They want, to the extent it can be given, a promise that if noted discrepancies are corrected, they will meet regulatory and corpo-rate requirements. In almost all instances, they view the audit report as a direct reflection on their abilities and performance as managers of the facility's environmental affairs. As such, they want their facility presented in the best possible light.

Understandably, while they generally want to hear about any per-ceived weaknesses in their environmental management program, they want to avoid or minimize reporting any negative aspects of their program upward within the corporate organization. They expect a report that fairly and factually represents the environmental compliance situation at their facility. They do not want items reported out of context. They generally want to exclude from reports to senior management items that do not involve clear and direct deviations from formal requirements. Also, if only one or two of 25 or 50 instances were observed to be deficient, they often believe that the 25 or 50 is at least as important a number to include in the report as the one or two problems.

Facility management wants an opportunity to respond to, or correct, deficiencies noted in the report. They generally want easily correctable exceptions quickly corrected so they can be omitted or deleted from the report. Where omission is not consistent with program goals, they quite legitimately want the report to reflect that the discrepancy has been quickly corrected.

Also significant to facility management's view of the reporting process is the basic coverage of the audit program. The needs of facility management can be influenced by the audit program coverage. Where all similar facilities in a company receive similar audits on the same repeat frequency, it can be much easier for facility managers to accept findings as reported. Such a situation usually means that all facility managers are being audited against the same general criteria. The repetitive nature of the audit schedule also usually means that discrepancies once noted can be addressed so that they are eliminated and, therefore, avoided in subsequent audit reports. However, the needs and desires can be quite different in situations where the environmental audit program functions more as a spot-check across the corporation. Here, facility managers are more apt to be concerned about whether their operations are being singled out unfairly. They often object to being held accountable for criteria that they believe are not being "enforced" uniformly across the company.

In some instances, facility management also may view the audit report as an opportunity to get support for an item that they already believe (prior to the audit) needs attention. Thus, the report can be an impetus for initiating measures that facility management believes is important in more effectively carrying out its responsibilities.

The Legal Department

The legal department also has its own set of needs and desires regarding the audit reporting process. They want to know that the legal interests of the corporation are not being jeopardized in any sense. They recognize that the audit team may be exposed to potentially sensitive information and want to ensure that all such information is handled in a manner that is consistent with, and supportive of, the best interests of the corporation.

There are varying views within the legal community about how much and what kind of documentation should be prepared in the environmental audit report. In many companies, the audit program has been established as an ongoing management tool rather than part of a particular legal strategy. As such, legal departments believe that shortcomings noted during the audit must be reported to those with the line responsibility and authority to correct the situation. They also recognize that management prerogative must be allowed for in that many of the exceptions noted will not necessarily represent a single, clear-cut course of action. They will, however, generally insist that an appropriate management review be made and that all noted exceptions be documented. In order to close the audit reporting loop, they often want documentation to be developed about what corrective action was determined appropriate, who is assigned the responsibility for correcting the situation, and when it is likely to be corrected. Similarly, in situations where corrective action is deferred or not deemed appropriate, they desire similar documentation to be developed to counter the "smoking gun" liability, which arises where records show that the corporation knew of a problem, documented that they knew of it, and then apparently did nothing to correct the situation.

The legal department also has a need to safeguard their basic ability both to protect especially sensitive information and also to defend the corporation when necessary. They will undoubtedly want an input into the design of the reporting process and may want to have some sort of ongoing participation. We have found that the role of legal in the audit reporting process often evolves to one of commenting on the audit team's accuracy in interpreting regulations, once the reporting process is well established. However, corporate legal departments are likely to become comfortable with audit reports that contain explicit written descriptions of any significant audit exception only after they are convinced that such exceptions will be followed up carefully.

The Audit Team

The audit program and the members of the audit team need to communicate the nature, scope, and general limitations of the audit so that the report is not misunderstood to be a guarantee or warranty of present or future compliance. Environmental auditors also need to know first hand that all observations have been communicated to the appropriate level(s) within the organization. This can be especially important in situations where an ad hoc audit team is utilized and its members have little or no direct involvement in the audit program after completion of the audit. No audit team wants to find itself trying to explain to corporate management why, if a particular shortcoming (which perhaps later turned out to be especially significant) was noted during the audit, that situation was either omitted from the report or presented in such a circuitous or cryptic manner that it was overlooked or misunderstood by the report addressee. When an audit identifies discrepancies that may be symptomatic of a lack of attention to detail or of more significant problems, the audit team will generally want them reported even if they are fully corrected before the final report is issued (or even if corrected before the audit is completed). In such instances, they generally will acknowledge in the report to management that a particular situation has been corrected (or that they were told it has been corrected).

PURPOSES OF THE AUDIT REPORT

In general, audit reports seek to provide clear and appropriate disclosure of audit findings. Within this overall goal they have three basic purposes: (1) to provide management information, (2) to initiate corrective action, and (3) to provide documentation of the audit and its findings.

Providing Management Information

As noted in Chapter 2, the purpose of the environmental audit report is strongly linked to the overall objective of the audit program. If the purpose of a program is, for example, to provide assurance to management, the audit report needs to provide information to top management on the more significant findings noted during the audit. The report may also address recommendations to correct any deficiencies noted. Further still, the audit report may include good practices of the facility.

If the goal of the audit program is to assist the facility manager, the information provided in the audit report should be detailed enough to help the facility manager know precisely what was wrong and how to improve the situation. Such audits are typically conducted by those directly involved in environmental management. The development of recommendations and/or action plans for inclusion in the audit report may, therefore, be within the authority or responsibility of the auditors and the scope of the audit program. Audit reports of this nature can list findings in a variety of ways—items needing immediate attention, items needing eventual attention, and items requiring further study. Typically, good practices are also included.

Initiating Corrective Action

A second use of the audit report is to initiate corrective action. This happens by structuring the linkage between the audit process and corrective action so that a well-defined sequence of steps is followed from the time findings are identified. If the program is structured in this manner, the environmental audit report identifies deficiencies to facility and corporate management in much the same manner as a financial audit report, namely, stating what the auditor did and what was found. In some cases, audit reports include recommendations to correct deficiences. Typically, however, line management decides what, if any, corrective action needs to be taken. At that time, action steps are set in motion to correct the deficiencies found. Of course, if the report is addressed to top management, people at operating levels will likely be highly motivated to take the corrective action. In this situation the audit report provides a vehicle by which corporate management can monitor corrective action.

The process of initiating corrective action begins with two required inputs: (1) a set of performance standards (generally environmental regulations and/or corporate standards) used by the audit team in conducting the audit and (2) a set of performance results that describe situations observed by the team in the same terms as the performance standards. In evaluating their observations, the auditors compare the performance standards with the actual performance results. If the inputs are at variance (i.e., performance results are not consistent with performance standards), the discrepancy is potentially an item to be reported. Whether the discrepancy should be included in the report depends on the purpose and design of the program, the significance of the discrepancy, and the scope of the report. In many cases, performance standards and performance results are stated in the report so that the reader clearly understands the exceptions noted.

If the reporting component of the audit program is properly designed, the report will initiate a review of the audit findings and start the process for determining what corrective action is appropriate to make future performance results consistent with the performance standards. In many companies, the environmental auditor is removed from line authority. In these cases, the responsibility for initiating corrective actions generally falls within line management. Thus, it is important that the report clearly and accurately communicate discrepancies to those persons with authority to initiate and carry out corrective action.

Providing Documentation

The audit report provides different types of documentation. First, the report documents the fact that an audit was conducted and, supported by the audit working papers, provides specific details upon which the report's findings are based. Second, the report documents the audit teams' understanding of the environmental status of the facility. Finally, the report, combined with the action plan and corrective action which should follow, can help provide evidence that attention and care are being devoted to environmental matters. The auditor needs sufficient documentation of what the audit did and did not include, as well as of audit findings and observations.

DETERMINING FOR WHOM THE REPORT IS INTENDED

Given the varying needs of different individuals within the organization, the audit program manager is faced with deciding where and how the information is to be directed, and what criteria are to be used for sorting out information into various reporting channels. To assist the auditor in such decisions, a multiple or hierarchical reporting scheme is generally appropriate. An example of such a scheme is shown in Table 7-1. As Table 7-1 shows, the facility environmental staff and facility manager want to be apprised of all deficiencies noted. Some audit learnings (housekeeping, for example) may not require reporting beyond the facility manager. As

TABLE 7-1.   EXAMPLE OF A HIERARCHICAL REPORTING SCHEME

Who	What	When	How
Facility environmental staff	All deficiencies noted	When noted	Orally throughout audit
Facility manager	All deficiencies noted	Periodically and at exit meeting	Orally during audit; oral and/or written during exit meeting
Corporate/division environmental staff	Deficiencies not locally correctable	Conclusion of audit	Informal communication and audit report
Operating management	Most significant matters	Conclusion of audit and periodically thereafter	Audit report and periodic meetings
Corporate management	General program status	Periodically during the year	Orally and/or in writing through audit report and/or status reports

described above, operating management typically wants to know the most significant matters uncovered during an audit. And, corporate management generally wants periodic reports which provide assurance that problems are under control.

In order to identify the appropriate persons to receive the audit reports, you must look at the goals and objectives of the audit program. Table 7-2 provides an example of the various primary addressee options of the report. Most companies send the audit report to one primary addressee with informational copies provided to others who have a need to know. However, the audit report may include more than one primary addressee so long as there is not ambiguity about the actions to be taken by each of the addressees.

While all audit reports are distributed to the facility manager being audited, in most companies the audit report has a relatively wide distribution. Typically, all audit reports are distributed to corporate environmental affairs. Some reports may also be distributed to various levels of operating and line management, the legal department, and the members of the audit team.

WHAT TO INCLUDE IN THE AUDIT REPORT

The type of information and level of detail to be provided in an audit report depend on the problem(s) identified and the individual(s) who has

TABLE 7-2.   OPTIONS FOR REPORT DISTRIBUTION

Purpose of Audit Program	Purpose of Audit Report	Primary Addressee Options
Provide assurance to management	Provide information to management on the more significant findings	President Executive vice-president Division (or group) president
Determine compliance status	Notify management of the status of the facility's environmental compliance	Facility manager Operating management Corporate environmental affairs
Assist facility manager	Provide facility management with information on their status and what they should do	Facility manager

to be notified. Generally speaking, the facility environmental staff and the facility manager have to be notified of all deficiencies. Specific problems have to be brought to the attention of the individuals responsible for the activity in which the problem occurs. Top management generally wants to know the overall compliance status and the major exceptions found. Audit reports addressed to them often are limited to a factual description of the findings and exceptions. In addition to this information, facility management typically wants to know the details of the auditor's findings and observations (and in many instances, what specifically the auditor believes needs to be done to correct identified deficiencies).

Table 7-3 lists the basic elements and content options for an audit report. Many companies include at least one of the content options; some include several in their audit reports. Typically, all audit reports contain a background or introduction that lists the purpose and scope of the audit, and identifies the audit team leader, team members, and other key audit participants.

Beyond these common elements, the diversity in report content begins. Most audit reports include opinions on the overall compliance with regulations as well as compliance with corporate policies and procedures. Some audit reports may list all applicable federal, state, and local requirements. Some may list cost-effective operations that the facility has employed. Others include a detailed description of the facility and its history, an impression of the facility management's ability to handle environmental risks, and/or recommendations and action plans.

In Table 7-4 four alternative audit report formats are presented to illustrate the array of choices available in organizing the content of the report.

Options for Review of Draft Report

The objectives of the report review typically are to resolve any conflicts and to clarify further the report's findings. Report reviewers vary from company to company, depending on the nature of the report and the interest of management. Normally, however, report reviewers include corporate environmental affairs, facility management, legal staff, and each member of the audit team. The law department is often significantly involved in developing the overall reporting process, yet tends to restrict itself, once the reporting process has been established, to commenting on the audit team's accuracy in interpreting the regulations. Facility management's role typically is to clarify or resolve specific items in the report and to prepare an action plan or response to the audit report.

TABLE 7-3.   REPORT CONTENT

A. Basic Elements

Purpose
   of audit
   of report
Scope of audit
Audit team
Time period under review
Exceptions

B. Content Options

1. General Information

Federal, state, and local regulations
Description of the facility
Questionnaires from the audit
History of the facility

2. Audit Findings

Environmental management and control system deficiencies
Good practices of the facility
Listing of violations/citations
Cost-effective operations
Strengths of the facility's environmental program
Statement on the facility's ability to handle environmental
   risks
Permit status
Analytic procedures

3. Recommendations and Follow-Up

Environmental management and control system recommenda-
   tions
Other recommendations
Action plans
Areas where the facility can save money

Occasionally, exceptions noted during the audit and resolved during
this review process can be omitted from the final report. For example,
facility management may find reports or other items that they were
previously unable to locate for the audit team. Moreover, when the audit
program is designed primarily to serve as a vehicle to help facility
managers achieve or maintain compliance, corrected exceptions may be

TABLE 7-4. ALTERNATIVE AUDIT REPORTING FORMATS

Alternative 1	Alternative 3
I. Background	I. Introduction
A. Purpose of audit	A. Purpose of audit
B. Scope of audit	B. Scope of audit
C. Conduct of audit	C. Scope of report
D. Audit participants	D. History of plant
II. Regulatory Compliance	II. Compliance Issues
A. Air	A. Air
1. Findings	1. Regulatory compliance
2. Exceptions	a. Regulations applicable to facility
3. Recommendations	b. Findings
B. Water, and so on[a]	c. Exceptions
III. Corporate Policy and Procedure Compliance	2. Corporate policy compliance
A. Air	a. Policies applicable to facility
1. Findings	b. Findings
2. Exceptions	c. Exceptions
3. Recommendations	B. Water, and so on[a]
B. Water, and so on [a]	III. Recommendations
IV. General Observations	
A. Items of potential concern	
B. Good practices of the facility	

Alternative 2	Alternative 4
I. Executive Summary	I. Background
	A. Scope of audit
II. General	B. Conduct of audit
A. Conduct of audit	C. Audit participants
B. Scope of audit	
C. Audit participants	II. Audit Questionnaires
III. Audit Results	III. Audit Findings
A. Air	A. Regulatory
1. Regulatory: corrective	1. Air
a. Requirements	2. Water, and so on[a]
b. Findings	B. Corporate policies and procedures
2. Regulatory: noncorrective	1. Air
a. Requirements	2. Water, and so on [a]
b. Findings	
3. Nonregulatory	IV. Action Items
a. Requirement	
b. Findings	
B. Water, and so on[a]	

Source. "Environmental Audit Reporting: A Study of Current Practices," Center for Environmental Assurance, Arthur D. Little, Inc. Cambridge, MA, 1982.

[a]Additional sections, as appropriate, for solid waste, safety, health and product safety, depending on the scope of the audit.

deleted from the formal report. Conversely, if the program is to determine compliance status during the period under review, such exceptions (although corrected during the review process) remain as items in the report.

The report reviewers should have explicit directions from the audit program manager (or team leader) about the report review process. Such directions can be included in a letter accompanying the report that states how and when they should reply. The audit program manager (or audit team leader) considers all comments received on the draft report. Since he or she typically maintains control over this review process, he or she decides whether or not to change the report based on the comments received. If conflicts cannot be resolved during the review process, the audit manager typically goes up the line of authority until such conflicts are resolved.

EXAMPLES OF REPORTING APPROACHES

After a decision has been reached about the purpose of the audit report, who the report is for, and what to include in the report, an overall reporting process can be designed. Table 7-5 presents some actual audit reporting approaches of various audit programs.

TABLE 7-5. EXAMPLES OF AUDIT REPORTING APPROACHES

Company A

Program objective	Determine compliance status with federal, state, and local regulations, and company policies and procedures
Purpose of audit report	Notify line and corporate management of facility's compliance status; help facility management achieve compliance
Primary addressee	Facility manager
Report distribution	Group vice-president, division head of group being audited, audit team, facility manager
Report content	Federal, state, and local regulations; exceptions; good practices; and recommendations

TABLE 7-5. *(Continued)*

Report timing	30 days after audit
Report response	Required within 30 days; 60-day follow-up until deficiencies are corrected
Company B	
Program objective	Provide assurance to management; verify compliance with federal, state, and local regulations; and determine internal management systems are in place to assure compliance
Purpose of audit report	Provide information to top management on the more significant findings of the audit
Primary addressee	Division president
Report distribution	Division president, legal, corporate environmental affairs, general manager of audited group, facility manager, audit team
Report content	Exceptions; recommendations regarding environmental management systems
Report timing	Within one month of audit
Report response	Required within two months
Company C	
Program objective	Provide facility management with the environmental status of the facility
Purpose of audit report	To help facility manager by providing information on the facility's status regarding regulations and corporate policy
Primary addressee	Facility manager
Report distribution	Division operating management, corporate environmental affairs, facility manager, audit program manager
Report content	Exceptions; good practices; recommendations
Report timing	Prepared on-site
Report response	Required within 45 days; six-month action plan follow-up

DATA MANAGEMENT NEEDS AND OPPORTUNITIES

The information and data generated during an audit present the need for some type and form of audit record keeping and data compilation. Such data compilation can take a variety of forms, such as the overall program status reports discussed in Chapter 14. Some companies have formal and sophisticated systems for compiling data from each audit. Audit findings can be computerized to assess similar problems of other facilities and to analyze trends in individual facilities. Data can be analyzed and used to plot the improvements of a facility.

Some companies use computers to keep abreast of the facility's action plan and the timetables for completion, thus facilitating the audit team's follow-up activities.

Chapter 19 addresses audit records retention, and how a formal records retention policy can aid in the collection, storage, and ultimate disposal of audit record keeping.

CHAPTER 8

ACTION PLANNING AND FOLLOW-UP

Chapter 8 addresses the various approaches to action planning and follow-up commonly found in many audit programs. Action plans are the methods developed to correct deficiencies noted by the auditors. Follow-up is the means for ensuring that corrective action occurs. Here we discuss procedures for responding to the audit report, audit follow-up responsibility, and managing audit follow-up.

PROCEDURES FOR RESPONDING TO THE AUDIT REPORT

Audit follow-up procedures are essential to the overall effectiveness of the audit program. Identified deficiencies must be reviewed and a decision made regarding corrective action. Whenever corrective actions are prescribed, a means of follow-up should be established to ensure that the corrective action is taken.

The audit program manager should establish formal action planning and follow-up procedures and clearly communicate them throughout the organization. The procedures can be communicated orally (at the audit's exit interview) and in writing (incorporated in the written audit report).

Table 8-1 lists a number of issues and options to be considered when action planning and follow-up procedures are being prepared.

Action plans—whether prepared on-site at the end of the audit or in response to the written audit report—should address all deficiencies noted by the audit team. In many companies, action planning is the responsibility of the facility manager. The action plan should indicate what actions will be (or have been) taken to correct any unsatisfactory conditions, who is responsible for ensuring that action is taken, and what the deadline is. If, on occasion, facility management (and/or operating management) decides not to correct a deficiency, this should be clearly stated in the action plan along with the reason why. If the corrective action described in the plan is deferred to a future date, the plan should indicate the date when corrective action will be completed. While action

TABLE 8-1. ISSUES AND OPTIONS TO CONSIDER FOR ACTION PLANNING AND FOLLOW-UP

Who should prepare action plans?
Facility management?
Corporate/division staff?
Facility management and audit team?
Other?

When should such plans be prepared?
On-site?
Immediately following the audit?
After final report is issued?
If after report, what is the timing?

How should the plans be prepared?
As a written report?
As a memorandum?
In outline form?

Who should receive the plans?
Audit program manager?
Operating management?
Environmental affairs?

What actions are required by those who receive the plans?
What are the responsibilities of the auditors in action planning and follow-up?
Should action plans be included as part of the audit report or as a separate document?

plans come in different forms, Figure 8-1 presents one example of an action plan.

In many companies, action plans are issued by the facility manager and prepared with input from those facility personnel responsible for the function. Such plans may be submitted to operating management or the audit program manager (with copies to those who received the audit

FIGURE 8-1. Example of an action plan.
___

June 1, 1984

TO: J. Jones, Division Manager

cc: D. Williams, Audit Program Manager
             R. Smith, Audit Team Leader
             P. White, Corporate Environmental Affairs

FROM: P. Mitchell, Newtown Facility Manager

SUBJECT: Action Plan in Response to Environmental Audit of XYZ's Newtown
             Facility, April 24–27, 1984

___

The following is the facility's response to the exceptions noted during the April 24–27, 1984 environmental audit of XYZ's Newtown Facility. The five deficiencies noted are addressed with corresponding target dates for compliance and individual accountabilities assigned.

1.   The facility's spill response plans will be changed to include all elements specified in the corporate guidelines. This includes

     a.   New oil unloading procedures have been written and incorporated into the facility's operating practices.

     b.   The facility will be fenced upon approval of recommendations by the appropriate headquarters personnel.

     Our action plan for this project is

     ☐   Obtain Quotes for Fencing Facility
          Responsibility: B. Carr, Facility Engineer
          Target Date: 6/22/84

     ☐   Overall Project Completion
          Responsibility: S. Anderson, Engineering Manager
          Target Date: 8/23/84

report). In some companies, plans are prepared on-site by facility personnel with the help of the audit team. These plans are then incorporated as part of the formal written report. The plan is generally reviewed by the individual with responsibility for monitoring follow-up—often operating management. The purpose of this review is to make sure that the action steps described are reasonable and appropriate. In many companies, the audit team leader is included in this review process to ensure that the action steps proposed do, in fact, address the deficiency noted, and that there is no misunderstanding of the audit team's findings. The typical audit follow-up steps discussed above and their timing are presented in Figure 8-2.

AUDIT FOLLOW-UP RESPONSIBILITIES

Identifying and reporting an audit deficiency and preparing action plans to correct it do not signal the end of the overall audit process. Such action plans must be monitored by the group that has follow-up authority—generally either operating management, environmental affairs, or the auditors. In many companies, the audit program manager (or audit team leader) has some role, either primary or support, in action planning and

FIGURE 8-2.   Typical audit follow-up steps.

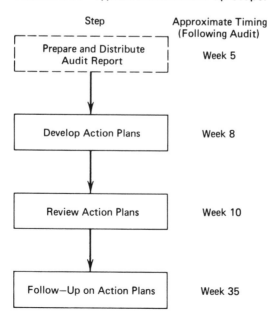

follow-up. A primary role is sometimes played by the auditor in companies where audits are repeated within a specific time (e.g., yearly). If an audit team is unlikely to return to the facility for some time, however, follow-up responsibility is typically assumed by operating management and the auditor's role is more one of support. This is also true where action plans are developed on-site and included in the written audit report.

Whoever is given the responsibility for follow-up, procedures should be established and communicated. In some companies, follow-up includes a monthly check on the status of action plans, 60-day checks, or six-month checks. Some companies further follow-up on action plans with a return visit to the facility by a representative of the audit program. Some commonly used approaches to audit follow-up are presented below.

Audit Follow-Up Alternatives

Issue	Alternative Approaches
Which audit follow-up methods should be used?	Oral status report Written status report Periodic check (phone call or letter) Repeat audit
What is the timing of audit follow-up?	Monthly Every 60 days Every six months Yearly
Who maintains follow-up responsibility?	Audit program manager (or audit team leader) Corporate environmental affairs Operating management

MANAGING AUDIT FOLLOW-UP

At a minimum, companies need to establish techniques and methods for managing audit follow-up. Follow-up schedules, program evaluation and review technique (PERT) charts, and reminder or "tickler" lists can be developed, maintained, and updated to insure routine and continued follow-up. PERT charts break down a project (action plan) into a series of

FIGURE 8-3. Example of an action plan status report.

XYZ COMPANY NEWTOWN FACILITY

Action Plan Status Report

April 30, 1984

Description of Action	Responsibility	Completion Date	Percent Completed	Comments
Update SPCC Plan	Smith	5/23/84	60%	Insufficient staffing; plan expected to be complete by 5/30 with approval by environmental affairs.

tasks and arrange them into a logical network. Such a technique provides the needed information to track events leading to the completion of an action plan.

Some companies with large audit staffs or a large schedule of facility audits have utilized computers to record and monitor audit follow-up. Action plans can be inputed into a computer with time schedules for both planned and actual action. Planned scheduled milestones remain constant; actual schedules may vary, thus, providing a tracking mechanism for the start and completion of corrective action. Computerized systems can produce status reports on action plans and facilitate tracking the progress of plans (noting percent completion). In smaller audit programs, this can be accomplished manually. Figure 8-3 (above) provides an example of such a report.

PART THREE

TECHNIQUES OF ENVIRONMENTAL AUDITING

CHAPTER 9

PRE-AUDIT ACTIVITIES AND AUDIT PLANNING

As described briefly in Chapter 3, an environmental audit involves a number of key steps, some of which are undertaken before the on-site review, some during the field work, and others after the facility audit has been completed. Chapter 9 focuses on those activities that are commonly carried out in advance of the audit; other audit activities are addressed in Chapters 10 through 14.

Planning for an environmental audit can be either a frustrating process or a productive exercise that will yield substantial benefits. The principal challenge for the audit team leader—particularly if the team leader is not assigned full time to environmental auditing—is to do all the things that need to be done despite the time and resource constraints. If one or more of the audit team members are "dedicated" part time or full time to auditing, less preparation time will be required. However, a substantial amount of the pre-audit work often involves interaction with individuals who have other job requirements (e.g., team members and facility manager) and who may not be able to devote the time desired. Whether the audit team is comprised of dedicated or rotating members, effective pre-audit planning will enable the auditors to do their job efficiently, have the audit run smoothly, and cause the facility minimal disruption.

The relative effort devoted to pre-audit activities varies from company to company and from audit to audit. For some, this preparation phase is more time consuming than either the on-site or post-audit phases. For others—particularly those who tend to be quite familiar with the audit process, the regulatory requirements, and perhaps the facility being audited—the pre-audit activities consume less time and effort.

While each environmental audit is likely to be somewhat different (both from audits of other facilities and from previous audits of the same facility), a number of initial activities are common to all. Table 9-1 lists these activities and summarizes the principal elements of each. Although the exact timing may vary substantially from company to company, these steps, described below, are commonly provided for in most established audit programs.

TABLE 9-1.   KEY PRE-AUDIT ACTIVITIES

Selecting facilities to be audited
 What facilities
 What functional areas
 What frequency

Scheduling the audit
 Schedule facility visit
 Select team members
 Confirm in writing

Gathering and reviewing background information
 Identify information requirements
 Conduct advance visit
 Review background information

Developing the audit plan
 Select priority topics
 Allocate team resources

Finalizing administrative details
 Confirm arrangements
 Select time and place for team members to meet
 Prepare appropriate items to take to facility

SELECTING FACILITIES TO BE AUDITED

The pre-audit activities generally begin with selection of the specific facilities to be audited. The selection process is often done annually, but may be modified during the year depending upon circumstances (such as scheduling problems and facility disruptions.)

Determining appropriate audit scheduling and review frequently depends upon the specific goals of the audit program and the number of facilities and functional areas included within the scope of the program. Facilities and functional areas can be selected by a number of methods; for example, random selection, perceived hazards, and the importance of the facility in terms of business considerations. An in-depth discussion of considerations involved in developing a typical schedule and audit cycle for your program is provided in Chapter 18.

SCHEDULING THE AUDIT

The initial arrangements relating to a facility audit include scheduling the visit, selecting the audit team members, and identifying the advance information needed. The specific types of arrangements that need to be made will depend on a number of factors that make each audit slightly different. One of the major considerations is whether the audit is the first at the facility, or more importantly, whether it is the first conducted by the corporation. The needed lead time is always greater for the first audit.

Scheduling the Facility Visit

The facility visit should be scheduled to minimize disruptions and inconveniences. The scheduling process generally begins with the audit team leader communicating to the facility manager that the facility has been selected for an environmental audit. This initial contact is usually followed up by a telephone call at least one month in advance of the audit to agree upon mutually acceptable dates for the audit. In selecting dates, the auditor's major considerations are to make sure the key facility personnel are available during the audit and to pick a week during which the facility is operating under normal conditions. After the arrangements have been agreed upon, it is common practice to send the facility manager a memorandum confirming those arrangements. Figure 9-1 provides an example of such a confirmation memorandum. Typically the

FIGURE 9-1. Confirmation memorandum.

MEMORANDUM

TO: Mr. John Smith, Newtown Facility Manager

cc: L. V. Dawson
            J. I. Jones
            P. O. Murray
            T. E. Thompson

FROM: M. J. Murphy, Corporate Audit Director

DATE: October 19, 1984

SUBJECT: Audit of XYZ Corporation's Newtown, New York Manufacturing Facility

---

This is to confirm the arrangements we agreed on for the upcoming environmental audit of the Newtown Facility during the week of November 26, 1984. During the audit, I will be accompanied by James I. Jones, Thomas E. Thompson, and Laura V. Dawson. We would like to begin the audit at 8:30 a.m. on November 27. As currently envisioned, we expect to conclude our audit during the afternoon of November 30 and will conduct an exit interview with you at that time to apprise you of all audit findings.

If convenient, we would like to have a pre-audit meeting with you and other facility personnel on Tuesday morning to describe briefly the purpose of the audit, discuss the approach we will take, and answer any questions you may have. Following this meeting, it would be helpful if we could receive a general overview of facility operations and the environmental organization. After this overview, we will begin our formal audit process.

As we discussed, the audit team will follow a detailed program designed to confirm that the air and water pollution control and solid and hazardous waste management practices of the facility are in compliance with regulatory requirements and company policy.

Please feel free to call if you have any questions or if scheduling conflicts arise.

---

memorandum includes such information as the scope, date, time, and place of the audit so that the right people and records are available at the time of the audit.

Selecting the Audit Team Members

About the time the team leader schedules the facility visit, he or she should select the team members. It is important to select the team members well in advance of the audit to confirm their availability, and to have replacements available in case of scheduling conflicts.

If the audit team members are dedicated full-time or part-time staff who have participated in an audit previously, little effort may be required to inform them of the audit details. On the other hand, rotating staff or dedicated team members on their first audit will need to learn certain details and procedures in advance of the audit. The groundwork for this training should be established when confirming their availability for the audit.

GATHERING AND REVIEWING BACKGROUND INFORMATION

This information gathering generally begins well in advance of the audit and can extend right up until the audit begins. The challenge is to identify the information desired and to review it in sufficient detail before the audit.

Identify Information Requirements

Soon after the schedule has been established, the audit team leader should identify the types of information needed or desired in advance of the audit. Some of this information may need to come from the facility; other types may be available from other sources.

It is important to recognize and be sensitive to the special needs of the facility personnel. Experience suggests that the facility manager and staff are generally very accommodating and will do what they can to be responsive to the team leader's requests. However, nothing will impede good relations faster than asking the facility manager to provide advance copies of documents and then not having the audit team review the documents in advance of the audit. This situation is magnified if one of the team members asks for a copy of a document that was provided to the team leader in advance of the audit. The following documents are often considered essential to review in advance of the audit.

Essential Information Requirements

Previous Audit Report
A copy of the previous audit report indicating discrepancies found

Regulatory Requirements
Copies of all applicable federal, state, and local regulations

Corporate Policies
Copies of all applicable corporate policies, procedures, and guidelines

Facility Information
Readily available information on the facility (process, layout, organization, etc.)

Examples of documents useful to review before the audit are noted below.

Optional Information Requirements

Facility Organization
Current facility organization chart annotated to illustrate line and staff responsibility for all environmental areas under review, and to identify key contact people

Facility Layout
Maps or diagrams to illustrate location of different operations and of environmental process and control system components

Permits
Listing of environmental permits, agencies responsible for review, and effective dates of each

Policy Manuals and Plans
Copy of current facility manuals, emergency plans, and so on, covering policies, operating procedures, and reporting requirements

Completed Questionnaires
Initial facility responses to specific questions asked during administration of questionnaires or other type of checklist

Selected Facility Reports
Copies of selected environmental status reports or other applicable reports sent to corporate headquarters

The auditor should consider obtaining copies of any or all of these documents that are available at headquarters rather than ask the facility manager. It is usually preferable not to bother the facility manager more than is absolutely necessary in advance of the audit. Obtaining a copy of the document directly from the facility rather than headquarters generally is appropriate only when it is important to ensure in advance of the audit that the facility has the complete up-to-date version.

Advance Visit to the Facility

One option available to audit program managers and team leaders is to visit the facility in advance of the audit. An advance visit can increase the effectiveness of the audit; however, it may not be appropriate in all situations.

There are a number of objectives for a pre-audit facility visit. First, the visit can be used to inform the facility manager about the audit program goals, objectives, and procedures. Second, it can provide the audit team with sufficient information to develop a basic understanding of the facility, the processes, and its environmental management systems. Third, the pre-audit visit provides a level of intelligence about the facility that allows the audit team to develop a more comprehensive audit plan prior to arrival at the site than might be developed without the benefit of the visit.

The costs and benefits of an advance visit should be identified and considered in the context of the goals and objectives of the audit program. The principal costs, summarized below, generally relate to the extra resources required as well as the "lost opportunity" on the part of team members who are not present; these are balanced against the benefits of extra intelligence gained from the pre-audit visit.

Principal Costs and Benefits of a Pre-Audit Visit

Costs
Extra resources (time and money) required by the audit team
Extra demands on facility management
Team members who are not present get second-hand information

Benefits
Early intelligence about facility operations
Facility management has a better idea of what is expected
Better able to "hit the ground running"
Build rapport in advance of audit
Minimize burden on facility (team rather than facility collects information)
Get first-hand information

A key factor in deciding whether an advance visit is appropriate is the proximity of the audit team leader to the facility being audited. If the team leader (and perhaps one or more of the team members) can visit the facility with a minimum expenditure of time and effort, an advance visit may be appropriate. If a significant distance separates the team leader from the facility, a pre-audit visit may not make sense.

Reviewing Background Information

Whether or not one decides to visit the facility in advance, the information gathered should be reviewed before on-site arrival. In order to effectively develop an audit plan tailored to each individual facility, the team leader, and team members as appropriate, should, at a minimum, review the essential information requirements described above.

This review of material prior to the audit should provide enough insight about the areas selected for the audit to enable the auditors to ask reasonably intelligent questions once the audit begins. The facility personnel do not expect the audit team to be experts in the facility's environmental programs at the outset, but a general familiarity with the regulatory requirements and their likely impact on the facility is highly desirable.

The result of this pre-audit review is generally a list of questions and issues to be used in developing the audit plan. Experience suggests that a comprehensive review of background information will minimize the risk of omitting an important step as the audit plan is developed and subsequently modified.

DEVELOPING THE AUDIT PLAN

As the necessary background information is gathered and reviewed, the audit team leader can begin to develop the audit plan. An audit plan is an outline of what steps need to be done, how each step is to be accomplished, who will do it, and in what sequence. Commonly, some form of audit protocol serves as the outline for the audit plan. What needs to be done in advance of the audit is to select the priority topics for review, modify and annotate the audit protocols or checklists as necessary, and make an initial allocation of audit team resources.

Priority topics are selected on the basis of the review of background information. The audit team will naturally want to focus its efforts on specific subject areas that are most likely to be of potential concern or pose a significant risk to the company. If a previous audit uncovered specific regulatory exceptions, a review of those areas would likely be a high priority. Likewise, known problems in a given area, such as excursions from the wastewater treatment facility or missing material safety data sheets, should be given a high priority.

The audit protocol or checklist should be modified and annotated after the regulations and other background information are reviewed. Specific protocol steps that clearly are not applicable should be marked as such. If there are special requirements for the facility that are not included in the protocol, they should be added. In essence, a protocol or checklist should be prepared that accurately identifies the areas to be audited in sufficient detail so that the auditors clearly understand what is required.

The final step in planning the audit is to make an initial cut at allocating the audit team resources. This involves matching the talent and expertise of the team members with specific tasks. At this pre-audit stage, some guesswork is almost always involved in allocating resources, for until the audit is actually under way, it is often difficult to know what is involved in completing the individual audit steps. Nevertheless, making an initial judgment about how best to match people with tasks allows further modification of the basic audit protocol to reflect the resources available to conduct the audit and enables the audit team members to begin the audit with an agreed-upon initial agenda and schedule.

During the development of the audit plan, a number of questions and issues will undoubtedly arise. The team leader should prepare a list of these for resolution at the preliminary meetings. Included in this list would be points that need to be clarified during the administration of the audit questionnaire, if one is used. (An in-depth discussion of audit protocols and questionnaires is provided in Chapter 10.)

FINALIZING ADMINISTRATIVE DETAILS

During the week or two before the audit, the team leader should take care of the final administrative details including:

Confirming arrangements with the facility manager.

Confirming availability of team members.

Setting a time and place for the team to meet before the audit begins.

Informing facility personnel of any documentation that needs to be made available for the beginning of the review.

It is helpful for the team leader to use some form of checklist to ensure that all of the necessary pre-audit tasks have been completed. The needs

TABLE 9-2.   EXAMPLE OF A PRE-AUDIT REMINDER LIST

	Date Completed	NA
Initial Arrangements		
Select facility for audit	_____	___
Determine scope of audit	_____	___
Schedule dates of audit	_____	___
Select team members	_____	___
Prepare list of advance information required	_____	___
Reviewing Information		
Federal regulations	_____	___
State regulations	_____	___
Local regulations	_____	___
Corporate policies	_____	___
Previous audit report	_____	___
Other: _____	_____	___
Planning the Audit		
Select priority topics	_____	___
Modify audit protocols as necessary	_____	___
Allocate audit team resources	_____	___
Final Arrangements		
Confirm availability of team members	_____	___
Confirm arrangements with facility manager	_____	___
Inform facility personnel of documentation that needs to be available	_____	___
Take necessary items to the facility	_____	___

of different companies will be satisfied only by lists tailored to their individual activities and methods. Thus, no general list will be appropriate for every audit program manager or team leader. Nevertheless, some form of pre-audit reminder list, such as that depicted in Table 9-2, can be extremely useful.

In addition to checking to make sure all the pre-audit activities have been done, the team leader should make sure all items, such as those listed below, needed during the audit are packed and ready.

Example of Items to Take to the Facility

Materials for presentation and discussion of audit objectives, scope, approach, reporting procedures, and so on

Applicable regulations and policies (that may not be available at the facility)

Copies of audit protocols and questionnaires for all team members (and others as appropriate)

Initial audit plan

Blank working papers

Other blank forms that may be needed:

    Form for listing documents included in working papers
    Form for documenting audit findings
    Close-out conference or exit interview form(s)
    Other

Miscellaneous materials (e.g., office supplies)

It is important to think through what kinds of documents or other articles may be needed (even such things as office supplies) in order to have necessary materials and to minimize disruptions to the facility.

CHAPTER 10

ENVIRONMENTAL AUDIT PROTOCOLS AND QUESTIONNAIRES

Most environmental audit programs use some form of written document to help guide the auditor through the on-site audit process. Such a guide, or protocol, adds consistency to the audit approach and can be used to help train the audit team. In addition, written audit guides are sometimes used to assist the facility in preparing for the audit.

Some companies prefer to use simple checklists or topical outlines that depend upon the auditors' knowledge of the regulatory and corporate requirements. Others prefer detailed guidelines that spell out the specific requirements and a step-by-step procedure for auditing against those requirements. Between these two extremes the choices vary from simple yes-or-no questionnaires to structured protocols that outline how the auditor should go about gathering information.

Chapter 10 discusses the basic purpose and uses of a written audit guide and presents a number of alternative formats. In particular, it examines two important audit guides—the audit protocol and the internal control questionnaire.

THE AUDIT PROTOCOL: PURPOSE AND USE

A variety of names are used to refer to documents that guide the auditor while conducting the audit, including audit protocol, audit work program, review program, checklist, and audit guide. Here, we will use the term *audit protocol*.

An audit protocol represents a plan of what the auditor is to do to accomplish the objectives of the audit. It is an important tool of the audit, since it not only serves as the auditor's guide to collecting evidence, but also as a record of the audit procedures completed by the team. As such, most audit protocols are either cross-referenced with the auditor's field notes or working papers or are formated in a manner that allows the field notes to be recorded directly on the protocol itself. Additionally, the completed audit protocol provides a record of the rationale for any deviation from plan or changes in audit procedures that occurred during the audit.

An audit protocol lists the step-by-step procedures that are to be followed during the audit to gain evidence about the facility's environmental practices. The protocol also provides the basis for assigning specific tasks to individual members of the audit team and for comparing what was accomplished with what was planned, and for summarizing and recording the work accomplished. In addition, a well-designed audit protocol can also be used to help train inexperienced auditors and thus reduce the amount of supervision required from the audit team leader.

Depending upon how the protocol is organized and formated, each audit step can be annotated on the audit protocol with specific working paper (the auditor's documentation) page reference, and initialed by the auditor as each procedure is performed. The completed audit protocol thus provides a record of the audit steps performed and documentation of the rationale for any changes or modifications made during the audit.

The audit protocol can be especially useful in building consistency into the audit, particularly in companies that use rotating audit teams or in programs with a large audit staff. In the larger programs, a comprehensive audit manual that includes internal standards and procedures can be particularly helpful in achieving such audit consistency.

However, an environmental audit protocol is not intended to be a rigid, absolute checklist that allows no deviation. Nor is it intended to substitute for sound, professional judgment on the part of the audit team. Rather, a protocol is a guidance tool designed to assist the environmental audit team in conducting a quality audit.

KEY ELEMENTS OF THE AUDIT PROTOCOL

While a variety of formats may be used for the audit protocol, depending on program goals and staffing approach, certain basic elements are common to most audit protocols.

Objectives

A description of the objectives of both the overall audit program and the audit protocol itself should be included (or referenced) in the audit protocol. Auditors must exercise judgment based on their knowledge and understanding of the overall audit program goals and any boundaries that they are expected to adhere to in modifying or deviating from the procedures in the company's environmental audit protocols.

Scope

An audit protocol should clearly describe the scope of the environmental audit so that the audit team has a firm understanding of what is to be included in the audit. For example, the scope of an audit may be limited to examination and verification of compliance with laws, regulations, and company policies; or the audit may include hazard identification and hazard control procedures and techniques. A single audit may include only selected aspects of a particular environmental area under review (e.g., only permits and reports under the NPDES/SPDES program) or may be defined more broadly to include several aspects of several different environmental areas. Similarly, the audit may be limited to only the facility's on-site activities; it may include on-site records and information of off-site activities such as an off-site disposer; or it may even include direct contact with off-site activities as well. The scope of the audit should also define the "period under review," (i.e., the time frame to be reviewed during the audit).

Subjects To Be Audited

The audit protocol identifies the subjects or topics within the audit scope that are to be reviewed or examined during the audit.

Audit Procedures

A comprehensive protocol not only lists the topics to be audited but also identifies the step-by-step procedures to be followed.

ALTERNATIVE APPROACHES AND COMMON PRACTICES

The audit protocol or guide can be organized and formated in a variety of ways. Six key alternatives, briefly highlighted in Table 10-1, are described in the sections that follow.

1. *Basic Protocol.* A basic audit protocol organizes audit procedures into a general sequence of audit steps and provides space for identifying team assignments, for brief comments and notations, and for cross-referencing with detailed working papers prepared by the team. Figure 10-1 illustrates one page from an audit protocol for solid and hazardous waste.

Common in many other types of auditing, the basic protocol is becoming widely used within many programs within the environmental auditing community as well. Among its advantages are its flexibility as an audit planning tool, its usefulness in focusing audit resources and organizing the working papers, and its applicability in evaluating and critiquing the audit process and results.

Audit protocols organize audit procedures into sequential steps and describe them in terms that specify the actions to be taken by the auditor. References to regulatory requirements or internal standards may be included depending on the level of training and familarity of the audit team. As illustrated in Figure 10-1, two columns are provided in addition to the list of audit steps. The comments column is used to identify which member(s) of the audit team has been assigned to carry out each audit step and is annotated to reflect any staffing modifications that are made as the audit progresses. The auditor in charge can also use this column to summarize any modifications to an audit step that were made during the audit. Reasons for the changes may also be summarized in the comments column. The working paper reference column is used to cross-reference pages in the working papers that provide the details of how the audit procedure was performed and results achieved for each audit step. (Working papers are discussed in detail in Chapter 12.)

Audit protocols prepared in this format can be quite lengthy. Audit step 6 in Figure 10-1 illustrates only one example—labeling and dating of

TABLE 10-1. AUDIT GUIDES

Approach, Style, Format	Characteristics
Basic protocol	A sequenced set of instructions outlining or specifying what is to be reviewed or examined, how the examination is to be conducted, and what is to be documented in the audit working papers; usually formatted to provide for quick identification of audit team assignments, deviations from audit plan, and cross-referencing of audit steps with working papers
Topical outline	Listing of topics to be included in the audit; generally leaves the manner in which each topic is audited up to the experience and discretion of the auditor
Detailed audit guide	Materials to familiarize the auditor with the basic requirements and standards against which the audit is to be conducted. The audit guide places substantial emphasis on requirements; often is in flow chart format
Yes-or-no questionnaire	Translation of the basic requirements against which the audit is to be conducted into a series of yes-or-no questions. Format can be expanded to provide for notation of the basis from which the answer is derived (inquiry, observation, testing) and/or cross-referencing of audit working papers. Tends to result in a questionnaire or inquiry driven rather than review or examination driven audit
Open-ended question-naire	A questionnaire designed and formated to include both selected explanations of yes-or-no responses and open-ended quesions that cannot usually be answered with a yes or no. Focuses the audit on fact finding and recording
Scored questionnaire or rating sheet	A questionnaire where responses are scored against criteria developed for that purpose resulting in either a numerical score or a satisfactory or unsatisfactory result. Tends to shift the audit team from review and fact finding to a grading role

FIGURE 10-1. Example of a solid and hazardous waste audit protocol.

	Auditor(s) Comments	Working Paper Reference (List Page Numbers)
Audit Steps		
1. Examine all relevant documents prior to visit to determine scope.		
Facility layout Regulations—federal, state, local Applicable permits Policies and procedures—corporate, facility Operating manuals Paper and report flow On-site systems for handling, storage, and disposal of solid and hazardous wastes		
2. Document the scope of the audit in the working papers.		
Time period under review Compliance auditing versus hazard auditing		
3. Develop an understanding of the facility's internal management controls through completion of the questionnaire or through other means such as discussions with facility manager, environmental manager, and so on.		
4. Document your understanding of the facility's internal control in flow chart or narrative form, showing responsibilities for action and record keeping for solid and hazardous wastes.		
5. Tour the sites to be audited to reconfirm and better understand the location's internal controls.		
6. Testing		
Identify different types of tests and describe in working papers. Prepare a schedule of key tests. Test selected transactions to confirm paper flow. Document in working papers. Develop a confirmation testing plan designed to verify compliance. Test selected transactions for compliance. Obtain instructions for labeling and dating wastes received into storage. Inspect labels and dates on a selected sample of wastes in storage. Document agreement between inventory records and wastes in storage. Record results in working papers.		

wastes in storage—of the tests that may be performed. In a comprehensive audit protocol, step 6 would include a listing and description of the tests considered necessary to verify compliance in the solid and hazardous waste area. Such a protocol could include several audit procedures for dozens of topic areas (e.g., emergency plans, classifying wastes, record keeping requirements, on-site treatment and disposal, off-site treatment and disposal, and transportation.

Although the audit protocol is common to many auditing efforts, it is far from universal. Some companies use checklists or guidelines to focus on specific items of laws, regulations, corporate policies, and management systems. Others use topical outlines to guide the auditor. Yet others use one of several types of questionnaires to guide their audit teams. Some even use a combination of two or more of these basic approaches and a flow diagram to assist the auditor in determining requirements.

2. *Topical Outline.*   A topical outline lists the topics to be covered during an environmental audit. It serves as a checklist by identifying the subjects to be included in the audit but, generally speaking, does not specify the precise procedures or manner in which each topic is to be reviewed. Thus, most topical outlines rely to a great extent on the experience and judgment of the auditor. Figure 10-2 provides an example of a topical outline.

3. *Detailed Guide.*   A detailed guide is intended to familiarize the audit team members with the basic requirements and standards against which the audit is to be conducted. Many detailed guides summarize both the basic thrust of a regulatory requirement and general industry responsibilities under the requirement. As illustrated in Figure 10-3, a flow diagram or other method may be used to assist the auditor in determining the applicability and implications of the particular requirement for the facility undergoing audit.

4. *Yes-or-No Questionnaires.*   A yes-or-no questionnaire is often used as the primary tool for gaining information about environmental performance. (See Figure 10-4.) Such questionnaires tend to be long and detailed; often they have been designed to incorporate virtually every regulatory provision into a question. As such, they may require relatively little background knowledge of the audit elements. When an audit is based primarily on an extensive questionnaire, the audit tends to be an inquiry-driven rather than review or examination-driven process.

Figure 10-5 is an example of an expanded yes-or-no questionnaire. The addition of an extra column provides a way of quickly recording the basis

FIGURE 10-2. Example of a topical outline.

Permits for
    Untreated wastewater discharges to surface waters
    Treated wastewater discharges to surface waters
    On-site disposal
    Discharge to POTW
    Other water permits (including NPDES)

Regulatory agency reports on facility expansion and production increases and process modifications.

Storm sewers conveying process wastewater or storm water run-off.

Cooling water and cooling water blow down.

Maintenance and calibration of
    Composite samplers
    Effluent flow measuring devices
    In-place monitoring devices
    Control devices

Laboratory analytical procedures for analysis of pollutants in compliance with 40 CFR, Part 136.

Discharges of pollutants from point sources.

Analysis of incoming water for pollutants controlled in facility permits.

Toxic pollutants, 40 CFR, Part 129

Laboratory analytical procedures complying with 40 CFR, Part 136.

Compliance with current and prospective EPA effluent guidelines.

Analysis of facility effluents for 129 priority pollutants.

Oil spill prevention, control, and countermeasures
    SPCC plan
    Arrangements for requesting assistance
    Underground metallic storage tanks
    Buried piping installations
    Mobile or portable storage tanks
    Tank loading and unloading procedures
    Spill prevention inspections and examinations

Groundwater

FIGURE 10-3. Example of a detailed audit guide.

Topic	Spill Prevention Control and Countermeasures (SPCC)

Summary and
Purpose: EPA requires that an SPCC plan be prepared and implemented by
any facility that could reasonably be expected to spill or discharge oil
into the waters of the U.S. (40 CFR, Part 112). These regulations
affect any facility that stores more than 1320 gallons of "oil-type"
material. An SPCC plan for the facility including information procedures relating to the prevention of spills and unauthorized discharges of oil and hazardous susbstances, must be maintained at the
plant and available for review by EPA or the Coast Guard in the
event of a spill.

Industry
Responsibilities: An SPCC plan is required for oil spills and should be prepared in
accordance with good engineering practices, with the full approval of
management at a level with authority to commit necessary resources.
The plan must designate what should be done for each spill. Facility
operators are responsible for training their personnel regarding the
SPCC plan. It is also desirable to have spill control plans for all
hazardous substances.

Required
Plans:

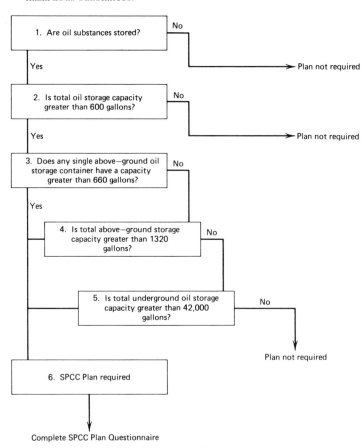

FIGURE 10-4. Example of a basic yes-or-no questionnaire.

Does the facility treat or dispose of any hazardous wastes, including containers, by means of incineration?   Yes____ No____

	Answer		
	Yes	No	N/A

1. Does the facility analyze wastes before burning them for the first time to establish proper operating conditions?   ____ ____ ____

   If yes, does the analysis include at least the following?

   (a) Heating value of the wastes   ____ ____ ____

   (b) Halogen and sulfur content   ____ ____ ____

   (c) Concentrations of lead and mercury in the wastes   ____ ____ ____

2. Are waste analysis results made part of the operating record of the facility?   ____ ____ ____

3. Does the facility incinerator have instruments that relate to the following aspects of combustion and emission control?   ____ ____ ____

   (a) Measurement of waste feed rate   ____ ____ ____

   (b) Measurement of auxiliary fuel flow rate   ____ ____ ____

   (c) Measurement of air flow rate   ____ ____ ____

   (d) Measurement of combustion temperature   ____ ____ ____

   (e) Measurement of scrubber flow   ____ ____ ____

   (f) Measurement of scrubber pH   ____ ____ ____

   (g) Pressure measurements throughout system   ____ ____ ____

4. Are the above instruments monitored at least every 15 minutes when hazardous wastes are being incinerated?   ____ ____ ____

100 TECHNIQUES OF ENVIRONMENTAL AUDITING

for the auditor's answer. Notation of the general technique used to obtain the response can be helpful in understanding and evaluating the audit. Similarly, the addition of a "working paper reference" column can provide space for recording short amplifying notes and cross-referencing more detailed notes on answers in the working papers.

5. *Open-Ended Questionnaire.*   An open-ended or expanded-answer questionnaire allows for more in-depth responses to specific questions. (Figure 10-6 provides an example of an expanded-answer questionnaire.) As is the case with most questionnaires, the format relies mostly on inquiry to gather evidence. Thus, the approach generally falls short of asking the audit team actually to verify compliance or confirm that adequate systems are in place.

This type of questionnaire does focus on obtaining factual and explanatory responses instead of, or in addition to, the yes-or-no answer. However, with open-ended questions, the auditor's field notes often tend to focus on a full response to the question without much notation (or in some cases, even awareness!) of the basis for the answer. For example, we have frequently found that auditors using this approach may be unaware whether the answer they have obtained refers to typical situations, the range of actual situations over some period, or the actual situation at the time of the audit. Additionally, when using an open-ended questionnaire, the auditor is encouraged to conduct the audit element by element, rather than to seek an overall understanding of the systems in place before gathering evidence. An open-ended questionnaire can provide a useful summary of environmental conditions as understood by those completing the questionnaire.

6. *Scored Questionnaire.*   This type of questionnaire attempts to measure environmental performance by rating each relevant activity against a detailed template, resulting in either a numerical score or a satisfactory or unsatisfactory rating. One example of a scored questionnaire is provided in Figure 10-7. As is the case with other questionnaire formats, inquiry tends to be the dominant method of gathering evidence.

A common criticism of scored questionnaires or rating sheets is that they can imply that some number of discrepancies is acceptable. There is a natural tendency to ask What is a "passing score"? whenever a scored questionnaire is used. Moreover, the approach attempts to rate all environmental activities numerically, even though some do not naturally lend themselves to numerical or other simple ratings. A strength is that a scored questionnaire may allow a facility to evaluate its own program in a manner consistent with the evaluation format used by the audit team.

FIGURE 10-5. Example of an expanded yes-or-no questionnaire.

Does the facility treat or dispose of any hazardous wastes,
including containers, by means of incineration?   Yes____   No____

Section A: RCRA Requirements—4O CFR Parts 264; 265, Subpart 0 (May 19, 1980)

	Answer			Answer Based on			Working Paper Ref.
	Yes	No	N/A	Inq	Obs	Test	
1. Does the facility analyze wastes before burning them for the first time to establish proper operating conditions?	——	——	——	——	——	——	
If yes, does the analysis include at least the following?							
(a) Heating value of the wastes	——	——	——	——	——	——	
(b) Halogen and sulfur content	——	——	——	——	——	——	
(c) Concentrations of lead and mercury in the wastes	——	——	——	——	——	——	
2. Are waste analysis results made part of the operating record of the facility?	——	——	——	——	——	——	
3. Does the facility incinerator have instruments that relate to the following aspects of combustion and emission control?							
(a) Measurement of waste feed rate	——	——	——	——	——	——	
(b) Measurement of auxiliary fuel flow rate	——	——	——	——	——	——	
(c) Measurement of air flow rate	——	——	——	——	——	——	
(d) Measurement of combustion temperature	——	——	——	——	——	——	
(e) Measurement of scrubber flow	——	——	——	——	——	——	
(f) Measurement of scrubber pH	——	——	——	——	——	——	
(g) Pressure measurements through-out system	——	——	——	——	——	——	
4. Are the above instruments monitored at least every 15 minutes when hazard-ous wastes are being incinerated?	——	——	——	——	——	——	

FIGURE 10-6. Example of an expanded questionnaire.

Part C: Solid and Hazardous Waste

2a. Are there known chemical burial sites within the facility boundaries? If yes, describe and discuss.

_____

_____

_____

_____

_____

_____

2.b. Is there any indication of groundwater contamination resulting from our operations? Contamination of storm water run-off?

_____

_____

_____

_____

_____

_____

DESCRIBING AUDIT PROCEDURES

Many audit protocols contain procedures that both guide the auditors on *what* to do during the audit and instruct them *how* to do it. As an example of guidance about what is to be done, an air pollution protocol might include the following audit step:

1. Review the facility's monitoring systems and note whether the monitoring requirements, as outlined in 40CFR Part 60.13 and applicable state and local regulations, have been adhered to. This review should include the following:

    (a) Determine whether monitoring systems and devices were installed and operational prior to conducting performance tests.

FIGURE 10-7. Example of a scored questionnaire.

Rating Worksheet

Waste Collection and Disposal

1. What solid wastes are produced as by-products of your processes?

_____

_____

2. Are solid wastes put in appropriate containers? _____

3. Are recycling programs effective? _____

4. Are liquid wastes disposed of effectively? _____

5. Are airborne wastes effectively controlled? _____

6. Do harmful wastes (e.g., radioactive materials and acids) receive special and proper handling and disposal? _____

Location _____

Rating _____

Rater (Name) _____

Date _____

Comments _____

(b) Confirm whether performance specifications, as noted in Appendix B of Title 40, Part 60, and applicable state and local regulations, are adhered to.

(c) Assess the adequacy of any alternative monitoring requirements used in the absence of a continuous monitoring system and document your assessment.

As to guidance with respect to how an audit procedure is to be carried out, a protocol can be explicit in telling the auditor to take such actions as

Develop an Understanding. For example, of the programs in place for hazardous materials disposal through reviewing paper flow procedures, examining record keeping systems, and identifying key controls.

Confirm. For example, that all emission sources for which permits are required are covered by permits through touring the plant and reviewing process flow diagrams.

Examine. For example, a random sample of the facility's quarterly monitoring reports submitted to federal, state, and local regulatory agencies, noting whether the information supplied complies with applicable requirements.

Review. For example, through a selected sample of maintenance department records that monitoring devices have been properly calibrated.

Verify. For example, through a sample of operating certificates and registrations relating to emissions from the facility that they are accurate and bear appropriate signatures.

Chapter 11 discusses in greater detail how to use these various audit procedures during field work.

PREPARING THE PROTOCOL

The audit protocol should be prepared early in the audit program development process, since it provides the structure of the individual audit. In preparing the protocols, the audit program manager should not only identify and outline the various environmental topics to be included in the audit but also specify the nature of the examination and the depth of review desired for each topic or subject to be covered. Typically, audit programs have a basic protocol for each functional area (e.g., air pollution control and water pollution control) that can be modified or tailored for each individual audit. The protocol should be flexible enough to allow for changes, but complete enough to guide the auditor throughout the audit.

In choosing among the basic protocol, topical outline, and questionnaire formats, the program manager should consider the amount and type of guidance needed by the audit team. Topical outlines usually require auditors to have significant knowledge about the regulatory requirements and company procedures. Questionnaires, especially extremely long questionnaires, frequently are completed mechanically without much thoughtful consideration, judgment, or verification. As a result, the program manager should design a process that encourages a scope of inquiry and evaluation consistent with the defined program goals.

In preparing the audit protocol, the audit program manager should consider which type of protocol is appropriate. The audit protocol can be either a "maximum program," which lists all possible procedures the auditor can perform, or a "minimum program," which specifies *required* audit procedures and can be modified by the auditor by the addition of audit procedures. Such modifications can be made for a variety of reasons including the nonapplicability of certain audit procedures, regulations, or policies; unique environmental hazards or risks that may be present at a facility; or operations and processes that are unique to a particular facility. The auditor may also modify the audit protocol according to the degree of reliance placed on the facility's internal control system.

Whether a minimum or maximum protocol is selected, a number of steps should be taken in developing the protocol.

1. Decide the scope of the audit.
    (a) Identify the functional areas to be covered (air, water, solid and hazardous waste, etc.).
    (b) Identify specific topics within each functional area (e.g., for water pollution control, topics might include permits, spill prevention, etc.).
    (c) Distinguish between on-site activities (e.g., plant operations) and off-site activities (e.g., hazardous waste transportation).
    (d) List your selection of audit topics.
2. For each topic selected in Step 1, identify and list the regulatory requirements (federal, state, and local) and corporate and facility policies and procedures applicable.
3. Determine and identify the depth of review for each topic selected above. For example, should the auditor examine *all* permits or a selected *sample* of permits; review records of *all* shipments of hazardous waste or a *sample* of shipments.
4. Determine the type and level of audit techniques (e.g., inquiry, observation, testing) to use for each topic selected, paying particular attention to audit resources and time constraints. (Refer to Chapter 11 for discussion of audit techniques.) In a solid and hazardous waste audit, for example, a test may be performed to verify that wastes have been tested to establish whether they must be subject to RCRA rules and regulations. Responsibility for such testing and the testing procedures used may also be included in the protocol.

5. Prepare draft of protocol from Steps 1 through 4 above.
6. Have the draft audit protocol reviewed for accuracy and completeness.
7. Revise and complete audit protocol.

It is important that the audit protocols be kept up to date and reflect current audit techniques used and the relevant regulatory and corporate guidelines that are to be considered in an audit. Moreover, the protocols should be revised based on experience gained in using the protocol during individual audits.

THE INTERNAL CONTROL QUESTIONNAIRE

Many environmental audit programs supplement the audit protocol or guide with a questionnaire specially designed to assist in efficient collection of specific background information about the facility's internal environmental management systems. Such questionnaires are commonly known as *internal control questionnaires*.

Purpose and Use

Internal control questionnaires are a vehicle to assist in identifying and reviewing internal management procedures and systems. They are used to supplement the audit protocol. Internal audit questionnaires should enhance the audit protocol rather than duplicate or substitute for it.

A well-designed internal control questionnaire should allow a large amount of background information to be collected quickly and efficiently. It should assist in identifying audit items not applicable to a particular facility's environmental management system. It is sometimes administered in advance of an audit to assist the auditors (and often the facility) in audit planning and preparation. No matter when the internal control questionnaire is administered, selected tests should be conducted during the audit to confirm the accuracy of the answer.

Few, if any, environmental audits include a sufficiently detailed audit of all activities, operations, and transactions generally required for the discovery of intentionally hidden compliance violations or other irregularities. Detailed audits of all transactions are usually not economically feasible, nor will they necessarily disclose *all* irregularities that might exist. Instead, auditors rely on the internal system of management procedures to provide reasonable, but not absolute, assurance of compliance and risk management.

Types of Questions

Internal control questionnaires generally include two types of questions: (1) those aimed at identifying and understanding key elements of the facility's internal environmental management procedures and systems and (2) those aimed at identifying audit topics that are not applicable at a particular facility.

Examples of questions aimed at identifying and understanding key elements of a facility's internal environmental management system include

Are hazardous waste storage areas inspected regularly for leaks, corrosion, or other deterioration?

Does the facility maintain a record of the identity and location of stored wastes?

Does the facility maintain a file of five-day letters submitted in accordance with the provision of the NPDES permit?

Such questions may deal with any of a variety of aspects of the management system including design, operation, maintenance, record keeping, and internal reporting. They are generally worded so that a "yes" response indicates the presence of a control element while a "no" may indicate a potential weakness.

Examples of questions pertaining to the applicability of certain audit procedures include

Does the facility discharge into a publicly owned treatment works?

Does the facility have an NPDES permit?

Are solid and hazardous wastes treated at the facility by any of the following methods:
    Incineration?
    Other thermal treatment?
    Landfill?
    Land treatment, for example, land farming?
    Physical or chemical treatment?
    Biological treatment?
    Underground injection?

Such questions are usually worded so that a "no" response indicates a set of requirements and corresponding audit procedures that are not applicable for the particular facility.

Internal control questionnaires may be limited to questions pertaining to compliance if the goal of the program is to audit the compliance status. However, where risk management is the overall philosophy and the audit objective includes verification of the effectiveness of internal management systems and procedures, the internal control questionnaire may include many questions on risk management and other systems not directly related to current compliance. In either situation, comprehensive internal control questionnaires can be quite lengthy. They may contain hundreds of questions covering environmental management procedures, responsibilities, duties, and other practices. Typically, the questions are organized by environmental topics.

Preparing and Administering the Questionnaire

When the internal control questionnaire is being prepared, the audit protocol may be used as a guide in determining the types of information needed. It is important that the questionnaire include the items necessary for the auditor to gain an understanding of the operations and systems within the facility. As a result, the scope of the questionnaire should be broad. It is easy to answer specific questions as not applicable. It is unlikely that the facility personnel will remind the auditor of questions he or she failed to ask. During the audit, the questionnaire can thus be used in determining specific items for follow-up. The information gathered in the questionnaire should be documented in the audit work papers.

The auditor must have a clear understanding of what information is needed when preparing the questionnaire, and, of course, know the objective, scope, and focus of the audit.

Figure 10-8 illustrates some types of information required in a solid and hazardous waste audit questionnaire.

FIGURE 10-8. Example of a questionnaire as a preliminary tool.

Solid and Hazardous Waste Questionnaire

This audit questionnaire is intended for designing and conducting facility audits. It may require additions, revisions, or other modifications in order to meet the needs of your particular audit objectives, industrial setting, or other special circumstances.

	Yes	No	N/A

1. Confirmation of facility as a generator.

   a. Has the facility characterized all solid wastes generated to determine which are hazardous under RCRA? ___ ___ ___

   b. Does the facility produce any wastes classified as hazardous?

       Listed wastes ___ ___ ___

       Ignitable ___ ___ ___

       Corrosive ___ ___ ___

       Reactive ___ ___ ___

       EP toxic ___ ___ ___

       Other. Please explain _____

       _____

   c. Does the facility have an EPA-issued generator identification number? ___ ___ ___

2. Does the facility treat, store, or dispose of hazardous wastes on-site? ___ ___ ___

   a. Does the facility have a RCRA permit? ___ ___ ___

   b. Does the facility have a written waste analysis plan? ___ ___ ___

   c. Does the facility accept wastes from other facilities for storage, treatment, or disposal? ___ ___ ___

3. Are solid and hazardous wastes treated at the facility site by any of the following methods?

   a. Incineration ___ ___ ___

   b. Other thermal treatment ___ ___ ___

FIGURE 10-8. *(Continued)*

		Yes	No	N/A
c.	Landfill	——	——	——
d.	Land treatment (e.g., land farming)	——	——	——
e.	Physical or chemical treatment	——	——	——
f.	Biological treatment	——	——	——
g.	Underground injection	——	——	——

There are also a variety of ways to administer the internal control questionnaire. Many audit teams meet with facility personnel at the beginning of the audit, with the audit team leader administering the questionnaire to the appropriate facility personnel. In this way, the entire audit team gains a first-hand understanding of the facility's response and can make further inquiries as necessary. It also may be administered formally by the audit team leader to facility personnel in advance of the audit. Or, it may be sent to the facility for completion by facility personnel and returned to the audit team leader prior to the audit. Various trade-offs in time, expense, information exchange, and rapport building are involved in selecting the most appropriate method of administering the internal control questionnaire.

CHAPTER 11

FIELD WORK

Field work is the term used to describe the on-site activities of the auditor. These activities relate principally to gathering sufficient data to achieve the goals and objectives of the environmental audit program.

Field work techniques are the tools of the auditor. In some cases, the goals of the audit program will call for absolute verification of a particular condition or situation. In other cases, less rigorous review may be acceptable. Effective auditing requires selecting a field work technique appropriate not only to the situation but also the program's goals.

Chapter 11 discusses a general approach to field work and presents the basic methods for audit data gathering. In addition, several skills for effective data gathering are presented and specific examples provided. Chapter 11 concludes with a discussion of basic guidelines for evaluating field work.

A GENERAL APPROACH TO FIELD WORK

In Chapter 3, we briefly discussed the five basic steps included in the on-site activities. During the field work, the auditor addresses the first three of these steps. The steps are depicted in Figure 11-1 (presented earlier as Figure 3-1) and are described in detail below.

FIGURE 11-1. Basic steps in the typical audit process.

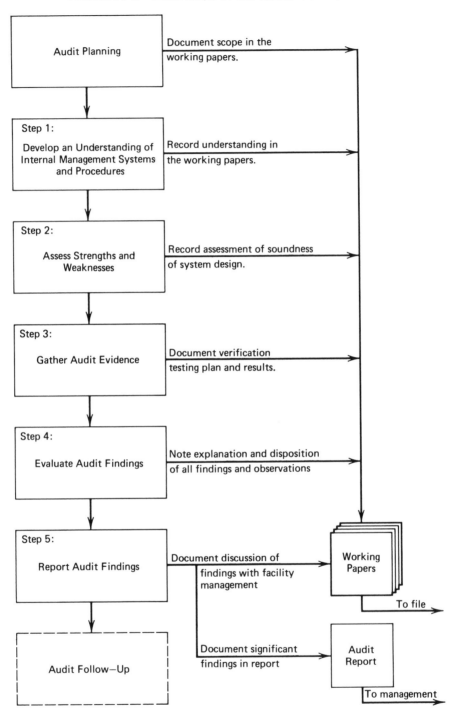

Step 1: Understand Facility's Internal Controls

The first step of the environmental audit is to develop an understanding of the facility's internal control system. *Internal control* refers to the actions taken within an organization to help the organization regulate and direct its activities. Most managed activities, by definition, have some sort of internal control. Every facility audited has an internal environmental management and control system. While this "management system" may not be very explicit or even thought of as a system by the facility staff, it will have at least some of the following:

Policies.   Information, directives, guidelines, and standards concerning operations and performance.

Procedures.   Instructions intended to ensure that operations are carried out as planned.

Practices.   Ways that people typically and routinely carry out operations.

Controls.   Checks and balances built into facility operations that may have an impact on environmental performance.

A management control system provides a framework for guiding, measuring, and evaluating performance. Management control systems can include both managerial controls and equipment controls. In addition, as shown in Figure 11-2, both the formality and the complexity of these systems can and do vary tremendously. In general, the more critical

FIGURE 11-2.   Range of management systems likely to be encountered.

Nonformal				Formal
←				→
A	B	C	D	E
Implicit policies and procedures	Formal policies; implicit procedures	Formal policies; limited written operating procedures	Formal policies; written operating procedures; limited control procedures	Formal policies; full written operating procedures; formal control procedures

to the organization a desired action or outcome becomes, the more appropriate or desirable a formal environmental management control system may be. Formal systems tend to be characterized by written policies (methods or courses of action designed to accomplish a purpose) and procedures (step-by-step instructions for describing actions to be taken), extensive documentation of standards (indicating what conditions should exist), timely control and measurement devices, formal behavior-altering devices (for bringing about required changes), and frequent reporting. Nonformal systems contain many of the above-described elements executed in the context of less formal, and often implicit, policies and procedures.

The goal of Step 1 is to develop a working understanding of how the facility intends to manage those activities that can influence environmental, health, and safety performance. This understanding of internal controls is important to the auditors both during and after the audit. It provides a benchmark by which the audit team can measure how accurately it understands the facility's management approach and processes. It also allows an auditor to assess the system's strengths and weaknesses and then to develop appropriate verification schemes.

In conducting an environmental audit, it is not necessary to interpret strictly what is or is not a management control system. In fact, experience suggests that both the environmental audit and the overall environmental management process can be enhanced if the auditor examines the management system, no matter how informal or unsophisticated, from the perspective of management control or internal controls.

Developing an understanding of the internal controls that are actually in place is seldom easy. Facility staff may not be accustomed to thinking about or discussing their procedures from a control system viewpoint. Even if they are, at least some of the actual procedures probably will differ from the prescribed practices. The job of an environmental auditor is to work through the "official" picture and develop an accurate picture of the way things really are.

An understanding of the internal control system is generally developed by several means.

1. *Review of Information.*   In the early stages, information is often obtained directly from the facility staff through background material provided to the audit team, discussions with managers and key staff, or briefings conducted by facility staff. Early review of background information such as an organization chart, can provide an initial overview of the basic responsibilities for internal control. Key documents that often are

reviewed to help understand internal controls include permits and applications; monitoring records and reports; safety records; SPCC, contingency, and emergency response plans; and operating manuals and log books.

2. *Questionnaire.* The auditor may also want to administer an "internal control questionnaire" early in this phase of the environmental audit process. As discussed in Chapter 10, the questionnaire is structured to get at the facility's management of basic environmental issues. For example, such a questionnaire might ask, Is a procedure in place to assure that proper Department of Transportation labels are on all drums prior to shipment? The auditor both records the yes-or-no answer and pays attention to the "soft" messages. In general, the questionnaire is designed so that yes answers indicate the existence of controls and no answers indicate a potential weakness in internal control. The questionnaire typically also includes yes-or-no questions designed to assist in determining the scope of the facility's activities with respect to the environmental audit. A simple example of this type of question is, Does the facility ship its wastes to an off-site disposal facility?

A well-designed and carefully executed questionnaire can be a powerful aid in developing an understanding of the facility's internal control system. However, the diversity of requirements and the relatively early stage of development of environmental, health, and safety management systems make it highly unlikely that a questionnaire alone will lead to a complete understanding.

3. *Plant Tour.* A plant tour that includes all process units, effluent discharge points, emissions sources, storage sites, maintenance shops, and administrative offices can be especially helpful in developing a basic understanding of physical relationships.

4. *Detailed Review.* With the briefings, questionnaires, and tour as background, the auditor will want to investigate the more detailed aspects of key control practices and procedures. This is likely to involve in-depth interviews, guided discussions, and often additional tours to specific sites.

As part of this detailed review, a limited amount of "verification testing" may sharpen the auditor's understanding of the system, check the reliability of the information received, or assist in identifying potential exceptions to the internal control system. While a limited amount of such testing may be appropriate, the data gathering methods used most in step 1 are inquiry and observation.

It is important for the audit team to record its understanding of the internal controls. This record may take the form of flow charts, narrative descriptions, or some combination that meets both the auditor's short-term needs (those relating to the subsequent steps in the audit process) and the long-term needs (such as establishing or defending the quality of the audit effort). Chapter 12 further discusses the importance of working papers and provides examples of various techniques for recording the auditor's understanding of the facility's internal controls.

Step 2: Evaluate Strengths and Weaknesses

Once an understanding of the facility's internal controls has been developed and recorded, the next basic step of the audit process is to evaluate the soundness of the design of those controls.

What constitutes satisfactory control? Some of the principles are listed below.

Satisfactory Control

Clearly defined responsibilities

Adequate system of authorizations

Division of duties

Trained and experienced personnel

Documentation

Protective measures

Internal verification

For example, consider clearly defined responsibilities. Here, as with the other characteristics of sound control, it is far easier to identify significant weaknesses than to determine adequacy. Where company policy calls for a Toxic Substances Control Act (TSCA) coordinator at each facility, it is reasonable to expect that the TSCA coordinator will be aware of his or her duties and those who need to interact with the coordinator will know to whom this duty is assigned. Should an auditor encounter a situation, as we once did, where two individuals in almost adjacent offices each thought that the other was the TSCA coordinator, it becomes obvious that the responsibility has not been clearly defined and that the control system is flawed.

Each of the seven characteristics listed above can require significant judgment on the part of the auditor. There simply are not many widely accepted standards, especially in the environmental, health, and safety management areas, that an auditor can use for comparison or as a guide to what is acceptable internal control. Nor are there any established principles or conventions about what constitutes an acceptable environmental audit. Accordingly, the auditor must keep referring to the goals and objectives of the audit program as well as to the corporation's basic environmental philosophy for guidance about what is satisfactory. While this is often not an easy task, it can be helpful in generating data and discussion to aid in the corporation's own development of internal standards for environmental management systems for its facilities.

The Step 2 evaluation of internal controls is important because, depending upon the outcome of the evaluation, the audit will proceed in one of two directions. In situations where the systems in place appear to be adequate, the auditor will test those systems during step 3 to confirm that the systems are actually in place and functioning as designed or intended. In those situations where the design of the system is assessed to be inadequate, further review or verification testing (to confirm that an inadequately designed system is, in fact, in place and functioning) is of no use. In this latter case, therefore, the evidence gathering focus of Step 3 must be entirely on specific compliance parameters rather than verification of the functioning of the system.

Step 3: Gather Audit Evidence

Having evaluated the facility's internal controls, the next step of the audit process is to gather evidence to identify and substantiate findings in accordance with audit objectives. Evidence can be defined as the information that forms the basis for the audit opinion. In carrying out Step 3, it is important for the auditor to keep in mind the purpose of the audit program. For example, evidence that supports a finding of "in compliance except for the following" must be much more extensive than that supporting findings of an audit with a goal of "identifying compliance problems." Verifying that the facility is in compliance suggests that the auditor will want to collect enough evidence to document that things were/are in compliance. That is, evidence that everything was in order; not just evidence associated with identified exceptions.

Three basic types of evidence can be gathered in an environmental audit—physical, documentary, and circumstantial—each of which has different implications for the auditor trying to develop accurate and conclusive audit findings. The first, physical evidence, simply stated, relates to something the auditor can see or touch. One example might be the physical sighting of a pH probe, an SO_2 monitor, or vehicle chocks. However, the presence of pollution control equipment or monitoring devices does not necessarily mean the equipment is appropriate or is functioning properly.

Documentary evidence pertains to something visible via a paper trail. RCRA manifests allow the auditor to see (in terms of documentation) those past hazardous waste shipments that have been manifested. Likewise, an accounts payable ledger would allow an auditor to see, "in documentary terms," those waste haulers who had been paid with company funds. As with other types of evidence, documentary evidence also has its limitations. A review of RCRA manifests does not tell the auditor about waste shipments that were not manifested. Moreover, documents can be generated or altered while preparing for an audit.

Circumstantial (indirect) evidence often is quite useful and informative in developing an overall impression. Examples could include the general order and neatness of files or records, supervisory qualifications and ability, and prevailing attitudes of operators and supervision.

The auditor goes about gathering and collecting evidence by following audit procedures that have been carefully devised by the audit program designer to be supportive of, and aid in accomplishing, audit program goals and objectives. In many environmental audit programs, the auditor places a heavy emphasis during Step 3 on verification testing, a form of field work described in the following section. Testing does not necessarily require the audit to involve effluent or emission sampling or chemical analysis. While these activities certainly fall under the umbrella of the term "testing," they are not often required by the environmental auditor.

The focus of the evidence-gathering step and the types of testing or verification that an environmental auditor will use depend greatly on the results of the first two steps of the basic audit process. Where internal control appears to be sound, the auditor will focus his or her efforts on testing the internal control system to verify that it does what it is supposed to do.

Keep in mind that the internal control system can include both managed and engineered controls. Accordingly, the auditor may perform "functional tests" aimed at verifying the performance of either the management system, the installed equipment, or both. In general, audit tests that include verification of the management system can be much more productive and fruitful in environmental auditing than tests that focus strictly on performance of the equipment at the time of the audit. For example, tests focusing on laboratory equipment could confirm that appropriate equipment is available and functioning properly at the time of the audit. Tests focusing on the entire internal control system could verify that standard methods are consistently used by the facility's laboratory; that properly trained, qualified, and supervised laboratory technicians conduct all analyses; that calibration, split sampling, and verification of results occur as appropriate; and that adequate documentation is routinely being developed.

As another example, consider a facility's internal control system for meeting NPDES permit monitoring requirements that includes a pH monitoring probe and recording device. Focusing on the equipment performance, the auditor can take prepared solutions and verify that the probe produces an accurate pH reading,which is recorded on the strip chart. Focusing on the management system, the auditor would examine records to verify that the various preventive maintenance, calibration, and testing activities required by the facility's management system had been done on time and were appropriately documented. In addition, the auditor could examine strip charts for selected periods to look for pH excursions or other irregularities.

Granted that this example may be a little extreme, the differences in results are obvious. In one situation, the auditor can get a good idea that the equipment is working at the time of the audit and even that it is likely to continue to work. In the other situation, the team member has determined what documentation exists to show that the internal control system does what it is intended to do.

Ideally, an auditor would like to use both these testing efforts. However, given the staffing and other resource realities of an environmental audit, the auditor cannot verify everything.

Steps 4 and 5: Evaluating and Reporting Findings

With the completion of Step 3, Gathering Audit Evidence, the field work is largely completed. However, many environmental audit programs accomplish portions of Steps 4 and 5, Evaluating and Reporting Findings, while on-site. Step 4 (Evaluating and Integrating Findings) and part of Step 5 (the exit interview portion of Reporting Audit Findings) are discussed in Chapter 13. The balance of Step 5 (the formal report) is covered in Chapter 14.

DATA GATHERING METHODS

During the on-site audit, each audit team member carries out a variety of procedures to determine the environmental compliance status of the facility. In the most basic sense, audit procedures are methods of acquiring audit evidence. While a more extensive classification of data gathering methods is possible—for example, observing, examining, questioning, analyzing, verifying, testing, investigating—the three basic approaches are inquiry, observation, and verification testing.

Inquiry

Inquiry is perhaps the most frequently used means of collecting environmental audit evidence. Here, the auditor asks facility personnel questions—both formally (e.g., via a questionnaire) and informally (e.g., through discussions). Inquiry often provides satisfactory explanations of unclear items in the records.

An auditor should consider the following factors in evaluating information gained through inquiry

The level of knowledge or skill of the individual questioned concerning the topic.

The objectivity of the questioned party. If management systems are well-developed, more weight generally may be given to individual responses.

The logic and reasonableness of the response. As an auditor gains familiarity with facility operations and organization, he or she becomes more adept at choosing the right person to question and evaluating the answer. The auditor also can determine whether it is consistent with other responses or information received.

While it is natural to rely heavily upon the responses of facility personnel, an auditor should recognize that this information may lack reliability. While the respondent may not intentionally deceive the auditor, it is human nature for facility managers and staff to want to describe facility practices in their best possible light. In addition, facility personnel may have inherent blind spots or biases of which they are simply unaware. The reliance placed upon evidence obtained through inquiry will vary, based on the factors discussed above, but heavier weight generally is accorded evidence generated independent of the auditee. An auditor should seek additional evidence whenever he or she judges a person's response to be uninformed, biased, or otherwise unreliable. In crucial matters, the auditor may also obtain additional evidence from independent sources.

Observation

Observation, or physical examination, is often one of the most reliable sources of audit evidence. Where specific operations are material to the environmental performance, it may be desirable for the auditor to observe such operations. In some cases, such as determining storm water run-off, it may be practical and desirable for the auditor to physically inspect the entire site. As noted earlier, however, inspection establishes or confirms physical existence, but not appropriateness or proper functioning.

Testing

Testing refers to the wide variety of verification activities that can be employed to increase confidence in the audit evidence and the facility's internal controls. Testing can be a very powerful technique in assisting the environmental auditor to achieve the objectives of the audit.

Auditors often develop evidence to form a conclusion through testing or sampling a portion of a whole collection (population) of items. The methods by which they select the sample can affect the validity of the sample and of the conclusions reached. It is important to minimize sampling bias and to obtain as representative a sample as possible. As a general rule, the auditor should maintain control of the sample selection. Independent records should be used whenever possible to develop the sample. No matter how carefully (or extensively) an auditor reviews a sample developed from a stack of training records provided by the facility coordinator, it is unlikely that the sample will provide any evidence of any untrained personnel. After all, the sample is of training records available to the facility coordinator and only reflects those who have been trained (or more precisely, those with completed training records). To gather evidence about the extent of training and training records, it would be more desirable to start with personnel department or payroll records and develop a sample of employees who should have been trained. Then, the training records could be reviewed to help determine whether each employee in the sample had been trained.

Samples selected by an auditor can be either judgmental or systematic. A sample developed largely on the basis of the auditor's judgment may be appropriate where the size or nature of the population make a systematic sample difficult or unreasonable to obtain. A systematic sample is one selected through the use of a systematic process chosen to represent the population that is being reviewed. Numerous methods are available to select a sample for review; however, no single method is a panacea for all situations. Among the methods available are:

Block Sampling. The objective is to analyze certain segments of records or areas of the facility. For example, if files were arranged alphabetically, in numerical order, or chronologically, one or more blocks (e.g., all the E's, records numbered 51 through 75, or January and June files) could be selected. While the block method is easy to use, it neglects entire segments of the population.

Random Sampling. The objective is to select items by chance. If properly done, each item in the population should have an equal chance of being selected and there should be no subjective determinations to bias the sample.

Stratification Sampling. The objective of stratification is to arrange items by categories (e.g., high versus low effluent volumes; new versus experienced employees; regular versus weekend or off-shift transactions) based on the auditor's judgment of risk. Higher risk categories then receive greater review and testing.

Regardless of the sampling method(s) used, it takes judgment both to evaluate the soundness of internal controls and to interpret the significance of the results obtained from verification testing.

A number of tests can be employed in an environmental audit. A few general examples of the many types available follow

Retracing Data. This method would uncover errors in recording original data. For example, hazardous waste shipments recorded on the annual report might be traced from shipping department records to the hazardous waste manifests to the annual report. Retracing usually proceeds forward from the original documents through the records. In this way, the auditor can check that all transactions were recorded.

Recomputation. Arithmetic calculations can be checked for accuracy. This would include, for example, recalculating the capacity of the diking around a storage tank.

Vouching. Vouching requires going to the original document that initiated the recording of the transaction. This document is then compared to the record, for example, to pH strip charts, operator logs, and NPDES reports. Vouching often goes backward. By starting with the records and moving back to the original document, the auditor can check that all recorded entries are supported.

Confirmation. Confirmation is written evidence from independent third parties. This procedure may be used where an auditor cannot physically observe a condition, such as an off-site disposal facility, operated by third parties.

SKILLS FOR EFFECTIVE DATA GATHERING

During the conduct of an audit, each team member will spend considerable time asking questions and engage in many informal discussions. Whether the auditor is gathering evidence through inquiry, observation, or testing, he or she will be interacting with facility personnel. To gain the maximum benefit from those interactions, the auditor should remember to adopt certain practices.

Whenever meetings or discussions are held with facility personnel, the auditor should take time to make the environment comfortable for everyone. If in an office or a conference room, rearrange the chairs if necessary so that all persons can sit in a comfortable position. It may be appropriate to close the door to eliminate outside noise and interference or to create an atmosphere of privacy. If the facility representative is busy or on the telephone when you arrive, ask if he or she wants five or ten more minutes to get things squared away before you begin—this is much better than being constantly interrupted during the interview. Be sure the facility representative knows your identity and why you are there. Describe briefly what you specifically want to learn from this individual.

Three key techniques for gaining useful information in interviews or discussions are

1. *Concreteness or Specificity of Response.* Focusing on the adequacy of the data included in a response and getting the interviewee to be as concrete as necessary.

2. *Respect.* Developing and maintaining rapport.

3. *Constructive Probing.* Focusing on resolving ambiguities or contradictions.

Concreteness

The most effective way to obtain specific and concrete responses is for the auditor to ask specific and concrete questions. Vague queries generally result in nonspecific responses that are seldom useful. The auditor must control the interviewing process both to elicit concrete answers and to limit the discussion to relevant issues.

Poor, fair, and good interviewer responses are illustrated below

Interviewing Example: Concreteness

Question: How are you tracking movement of hazardous waste from the plant to treatment, storage, and disposal sites?

Interviewee: Oh, that's a top priority here and we're doing all we can to comply with federal regulations.

Poor

> *Interviewer:* Fine, then you're checking to be sure your contractors are properly permitted or licensed and you're keeping good tabs on your manifest, and so on.
>
> (*Note:* Statements such as this, which essentially "put the answer in the interviewee's mouth," will ususally elicit agreement only.)

Fair

> *Interviewer:* Are your transporters properly permitted and do you have a control system to ensure that manifests are properly processed and returned to you?
>
> (*Note:* this type of question will ususally lead to a "yes" response and little or no additional information.)

Good

> *Interviewer:* How are you ensuring that your contractors are in .compliance, and how do you track your manifests?
>
> (*Note:* This will generally result in more detailed response.)

Respect

Respect is effectively expressed when the auditor focuses intently on understanding the interviewee's responses. There are few more direct communications of respect than the commitment to understand the interviewee. That is, one should focus on the *information* being given while deferring critical judgments about the respondent. Poor answers often do not indicate that the interviewee lacks the ability to respond adequately, but rather that he or she is anxious about the interview, or that the question is open to more than one interpretation. Helping the interviewee to clarify and/or deepen his or her responses communicates respect and interest and provides a vehicle for eliciting specific responses.

Poor, fair, and good auditor responses to show respect are illustrated below.

Interviewing Examples: Respect

Question: How are you tracking movement of hazardous waste from the plant to treatment, storage, and disposal sites?

Interviewee: Oh, that's a top priority here and we're doing all we can to comply with federal regulations.

Poor

> *Interviewer:* You will really have to be clearer than that if this audit is going to be of any use to us.
>
> (*Note:* This is likely to put the interviewee on the offensive and reduce the flow of information.)

Fair

> *Interviewer:* I really need to know more specifically what you're doing.
>
> (*Note:* This is likely to puzzle the interviewee since no help is provided in defining the "what.")

Good

> *Interviewer:* It's good to know that you're making a real effort in this area. Can you describe your procedures for tracking manifests?
>
> (*Note:* This is likely to result in a more specific answer and to help build rapport with the interviewee.)

Constructive Probing

Constructive probing is often necessary, especially when interviewees provide responses that are inconsistent or conflicting. When questioned about the apparent inconsistencies, respondents are usually able to explain them satisfactorily. It is important, though, that the auditor phrase inquiries to focus on the *data* rather than confronting the respondent; that is, the effect of the inquiry should not be to criticize the respondent for being inconsistent, but rather to enlist the help of the respondent in clarifying the information.

The following illustrates poor, fair, and good auditor constructive probing.

Interviewing Examples: Constructive Probing

Question: How are you tracking movement of hazardous waste from the plant to treatment, storage, and disposal sites?

Interviewee: I have my secretary call the contractor every so often to check on our manifests—we are very committed to complying with the federal regulations.

Poor

Interviewer: I see, then she's pretty conscientious about following up on those things.

(*Note:* this will usually lead to premature closure rather than the eliciting of more detailed information.)

Fair

Interviewer: How do you ensure that she follows up on all shipments?

(*Note:* This does not confront the questions of the basic adequacy or inadequacy of the control process.)

Good

Interviewer: I'm not sure about your overall process for tracking manifests and how your secretary's efforts fit in. Could you explain the process to me?

(*Note:* This restates the question in a more focused way and respectfully challenges the notion that the secretary's efforts are adequate.)

In ending the interview or discussion, it is often useful to ask a question such as, Is there anything else I should have asked you and haven't? The auditor might also ask if the facility representative feels that the information given offers a fair and accurate description of the facility's efforts. In any case, be sure to thank the interviewee for his or her time and reinforce that the interview has been useful and helpful.

These kinds of interviewing techniques are skills and, as with any skills, one needs practice.

GUIDELINES FOR EVALUATING FIELD WORK

As field work is completed, it is important to determine whether the information gathered by the auditor during field work is sufficient to support the objectives of the audit and the conclusions of the auditor. As mentioned above, evidence in the environmental auditing process can be defined as whatever influences the auditor's findings and opinions. As such, evidence should be relevant, free from bias, objective, and persuasive.

The first three properties define the competence of evidence. The last requirement, persuasiveness, refers to its sufficiency.

Relevance

Environmental audit evidence should produce a flow of logic from the auditor's discoveries to the assertions to be tested. Thus, confirmations of hazardous waste manifests could constitute evidence that the reviewed shipments were appropriately manifested in terms of the Environmental Protection Agency's manifest documentation requirements, but would not support the supposition that all hazardous waste shipments of the facility have, in fact, been recorded and documented. The concept of materiality is also tied to relevance. The auditor should focus his or her concern on verifying those environmental assertions with significant consequences.

Freedom from Bias

Evidence must be free from any influence that would make one decision more attractive than another and, therefore, exclude evidence which would support the alternative decision.

Bias can arise from the source of the evidence or from the auditor's choice of items to examine. The answers received when questioning management about their adherence to internal control procedures may be biased, because it would be in management's best interest to appear competent and efficient. If an auditor decided to examine a random sample of available records without first determining that available records represented all transactions, the sample might be biased. Similarly, a sample based on the shipping log could exclude off-hours or weekend shipments unless provisions had been made to ensure that all such shipments were later logged. Also, observations collected during a brief walk-around are likely to omit data points that are less accessible or out of the way and could, therefore, bias the evidence.

Objectivity

Although objectivity is close to freedom of bias, to the auditor it has a special meaning. The objective quality of evidence should lead two auditors examining the same evidence to reach the same conclusion. If, based on available evidence, two auditors reach different conclusions about a facility's compliance with particular requirements, the evidence lacks objectivity and, therefore, is unreliable for a decision.

Persuasiveness

Evidence is persuasive when it forces a conclusion to be drawn and when that same conclusion is reached by different people. For example, the evidence is very persuasive that life does not exist on the planet Mercury, less persuasive that it does not exist elsewhere in the universe. The persuasiveness may come from the volume of evidence, from the type of evidence, and from the source of evidence.

Evidence from some sources is more weighty than from others:

Independent sources outside an enterprise provide greater assurance of reliability than evidence from within.

Data and statements developed under satisfactory conditions of internal control are more reliable than those developed under unsatisfactory conditions.

An auditor's direct personal knowledge through physical examination, observation, computation, and inspection is more persuasive than evidence obtained indirectly.

CHAPTER 12

WORKING PAPERS AND AUDIT RECORD KEEPING

The working papers in many ways represent the core of the environmental audit—the documentation of the work performed, the techniques used, and the conclusions reached. While Chapter 11 discussed the nature and techniques of field work, Chapter 12 outlines the basic techniques and methodology used to record the procedures followed, observations made, discussions held, and conclusions reached by the audit team members during their field work.

PURPOSE OF WORKING PAPERS

Working papers help the auditor to achieve the audit objectives and provide reasonable assurance that an adequate audit was conducted consistent with program goals and objectives. Because the working papers document the information gathered by each audit team member during the audit, the information included should substantiate the conclusions reached about the areas of both compliance and noncompliance.

Working papers also can:

Supplement the audit protocols by providing audit planning details (such as the time allotted by the audit team to each audit step), and specific documentation of the internal control systems.

Provide the rationale for the auditors' approach to testing, a record of tests conducted, and the evidence accumulated.

Provide data that support the audit report and that may be useful in subsequent action planning and follow-up activities.

Provide a basis for review of the audit by the team leader, the audit program manager, or some other individual.

WORKING PAPER CONTENTS

The working papers should contain all information the auditor believes may be necessary to support the audit findings. It is essential, therefore, that the papers be complete, accurate, clearly indexed, neat, and legible. They should represent a full record of the work performed by the auditors.

Generally speaking, the environmental audit working papers contain three basic elements: (1) a description of the environmental management systems in place for managing various aspects of compliance, (2) a description of the specific audit methods or actions taken to complete each step of the protocol [including documentation of the tests conducted, sources(s) of information obtained, and evidence accumulated], and (3) a summary of the auditors' findings and observations along with the conclusions reached at each audit step.

Description of Management Systems

The first component of the working papers is a description of the internal systems in place for managing various aspects of environmental compliance. An auditor can describe management systems in several ways, ranging from narrative descriptions to elaborate flow charts. The selection of the device to be used in describing the management systems depends upon both the characteristics of the activity being reviewed and the individual preferences of the auditor. Whichever device is used, the approach taken should facilitate both clear note taking and subsequent review and audit of the management system design and implementation.

Various diagrammatic sketches can be used to describe the facility's internal management systems in the audit working papers. Flow charts can provide the basis not only for documenting the auditor's understanding of the management system but also for tracing, highlighting, documenting, and systematically evaluating the sequence of events and steps involved in any work process or procedure. Developing an explicit understanding of the sequence of events and steps involved can provide the basis for in-depth audits of practices versus plan.

Flow charts can provide an efficient way to document the auditors' understanding of the internal management systems. However, like other audit techniques, flow charting requires some skill development. Among the methods used to chart the work or information flow are:

Single-Column Work Flow Chart.   A simple schematic diagram that provides an overview of the sequence of operations or steps in relatively simple processes. (See Figure 12-1.)

Multi-Column Work Flow Chart.   A schematic flow diagram used to highlight interactions within complex processes or when many organizational functions are involved. (Figure 12-2 provides an example of a multicolumn work flow chart for off-site shipment of waste.)

Paper Flow Analysis.   A schematic diagram that focuses on flow charting the movement of information (especially paper and other documentation) rather than operational steps. (See Figure 12-3.)

Description of the Action Taken to Complete Each Step of the Audit Protocol

In addition to describing the management systems in place, the working papers should also describe the process the auditor used to gather information about a specific step in the audit protocol, the information and facts collected, and the sources of the information gathered. After stating the purpose of the audit step and the general understanding of the management systems in place, the auditor should clearly identify the rationale for the tests to be performed, the sample selected, the results of the tests, and the observations made. For example, if the protocol calls for a review of all air emission sources, the action taken might be a review of facility plans, a discussion with the facility engineer, a facility tour with the environmental coordinator, notation of all emission sources desired, and the comparison of noted sources with reviewed plans.

To substantiate that the auditor adhered to the audit protocol or

FIGURE 12-1. Example of a single-column flow chart.

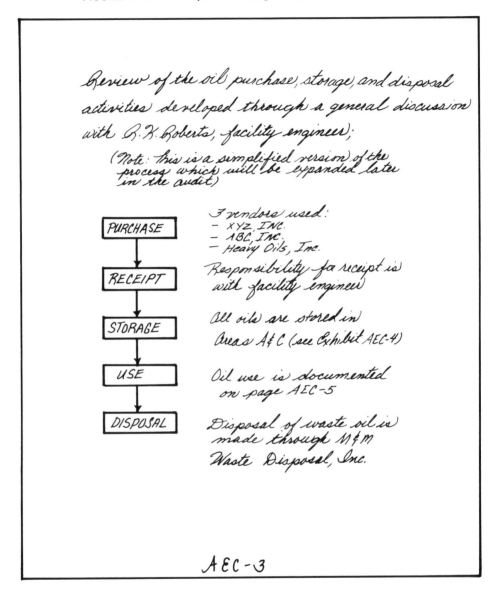

Review of the oil purchase, storage, and disposal activities developed through a general discussion with R. K. Roberts, facility engineer;

(Note: this is a simplified version of the process which will be expanded later in the audit)

PURCHASE

3 vendors used:
- XYZ, INC.
- ABC, INC.
- Heavy Oils, Inc.

RECEIPT

Responsibility for receipt is with facility engineer

STORAGE

All oils are stored in Areas A & C (see Exhibit AEC-4)

USE

Oil use is documented on page AEC-5

DISPOSAL

Disposal of waste oil is made through M & M Waste Disposal, Inc.

AEC-3

132

FIGURE 12-2. Example of a multi-column flow chart.

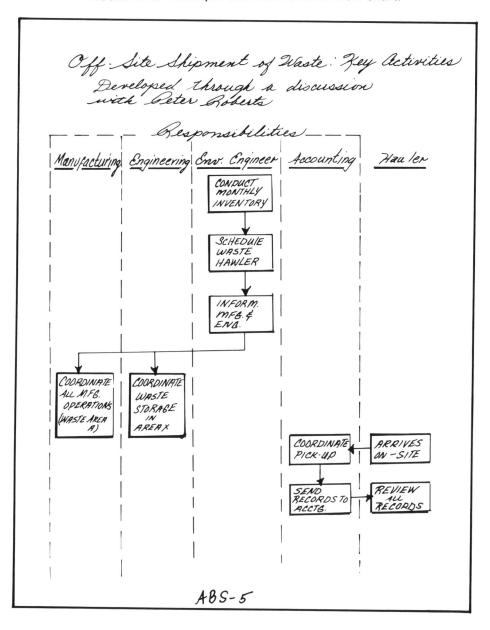

133

FIGURE 12-3. Example of a paper flow analysis.

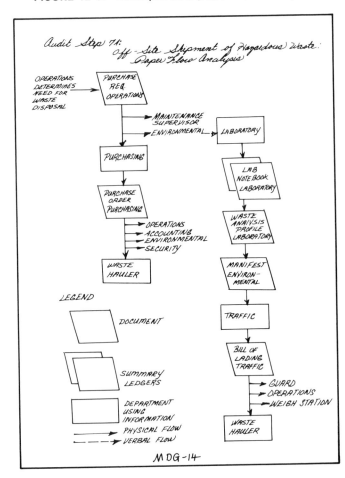

modified the procedures as required in conducting the audit, the working papers should provide detail about both the testing rationale used and the resulting evidence for each item in the protocol. For example, if the action taken was to tour the facility randomly inspecting the labels on hazardous waste containers, the results might be documented in the working papers by means of a table that lists the number of containers inspected, their location, and the information on each label. Similarly, if the action taken was to compare every other month's NPDES reported data with strip chart record data, the working papers would include a description of the sample (every other month), an explanation of the rationale, and a table illustrating the result of the comparison.

Summary and Conclusions

Under the sometimes intense time pressures and constraints of an environmental audit, the auditor is often tempted not to take all the time necessary to review the observations made and to summarize the open-ended items at various times during the audit. There is considerable value to taking a few minutes at selected time intervals—perhaps at least at the end of each day—to list tentative conclusions reached and open-ended items that need further attention.

This process of summarizing and drawing conclusions involves having each auditor review (1) the audit activities undertaken, (2) the results achieved, and (3) the notation recorded in his or her working papers. The results of this review are also noted in the working papers. Additionally, it can be useful for the team leader to conduct a similar group review of progress made and results achieved and record the results of the collective team discussion in the working papers.

These interim summaries and conclusions should be clearly labeled and identified in a separate section of the working papers. In addition, it is critical that the auditor go back to these lists later in the audit to cross off open-ended items as appropriate and to modify or confirm tentative conclusions so that the final working papers represent a cohesive, internally consistent package.

WORKING PAPER ORGANIZATION

One of the more difficult aspects of environmental auditing is to acquire the skill to develop comprehensive, organized, and coherent working papers *while* the audit is being conducted. There are two aspects to the organization of working papers: (1) how to organize them during the on-site review and (2) how to reorganize them after completion of the audit. A few individuals perfer to write—or at least rewrite—the working papers after the audit is over. However, working papers prepared after the audit often fall short of being the detailed documentation of audit steps undertaken, sampling rationales selected, results noted, and so on, that are desired. Hence, we suggest that the environmental auditor practice the self-discipline necessary to develop appropriate working notes during the audit.

On-Site Organization of the Working Papers

Four techniques are basic to developing thorough and well-organized working papers during the audit. First, write while conducting the audit, citing the protocol step, describing the actions taken to carry out that step, listing the persons interviewed and other sources of information, and describing the facts collected. Second, start a new page when moving to a new topic or new aspect of your investigation. Third, review working papers frequently during the audit and upon completion of the field work. Clarify and add to notations as appropriate. Finally, at the end of the audit, sit down and reorganize working papers. (Here, the frequent pagination suggested as the second step can help immensely.)

One important question in determining how to organize the audit working papers during the field work relates to how the auditor sorts through and arranges the often considerable volume of data and documents so that another person reviewing the papers can clearly understand the situation. There are two basic options regarding how to incorporate back-up documents into the working papers.

1. Insert back-up material directly after the page in the working papers where that document is first referred to.
2. Put all the documents together as a separate section of the working papers with an exhibit list summarizing the documents.

Each option has advantages and disadvantages. The first offers the convenience of referring to a document directly behind the page on which it is referred to, and the advantage of more coherent descriptions when reviewed by audit supervision or other reviewers. However, the insertion of several large documents can make the working papers very bulky and sometimes difficult to read. Moreover, insertion of documents in the working papers by each auditor can result in documenting duplicate copies of several key supporting documents (e.g., key permits and local procedures) within the working papers as several members of the audit team reference (and collect copies of) the same document.

A separate exhibit section keeps the auditor's comments section to a more easily manageable size, but separates the document from the page on which it is referred to.

Working Paper Organization After Completion of the Audit

Working papers should be organized in a way that is both easily understood by someone not familiar with the details of the audit and will minimize the opportunities for misinterpretation of the facts. To limit the access of others to the audit records, original working papers are commonly stored in a central file and not copied. Often, this requires that each team member complete the working papers *before* the team leaves the facility. Upon completion of the audit, the working papers are usually segregated between a "permanent" or "continuing" working paper file and the current year's audit papers. The permanent audit file includes a record of items of continuing interest, while the current file contains items unique to the most recent audit.

Permanent File

The permanent file typically contains those papers necessary or helpful for subsequent environmental audits of the same location. For example, the permanent file generally contains those items listed below.

Permanent File Contents

A listing of the facility's processes, production units, and so on.

Copies of all relevant regulatory permits and key regulatory correspondence.

Internal control questionnaires and flow charts.

Schedules of documentation and record keeping.

Copies of previous years' statements of findings and observations.

Current File

The current file contains the auditor's evidence and final decision and includes items such as those identified below.

Current File Contents

Notes of audit planning (including audit leader, team members and assignments, time budgets, rationale for any audit program modifications, etc.).

The completed copy of the audit protocol annotated with working paper references and auditor's initials.

The completed internal control questionnaire, results of compliance testing, and evaluation of internal control.

Schedules for compliance testing.

Descriptions of all transactional and functional tests conducted during the audit, as well as an explanation of sampling plans employed.

Documentation of all audit procedures and evidence obtained (including favorable as well as unfavorable evidence).

Notes on any conferences or meetings held with local facility management.

The working papers that comprise the current file should document a flow of logic from the assertions to be tested to the auditor's discoveries. They are frequently written in a narrative discussion format that describes the particular audit objectives, the step-by-step process used by the auditor, and the resulting evidence for each section of the audit protocol.

The current file working papers are typically completed on-site at the time of the audit and are maintained in their handwritten form. They often include extracts or photocopies of various documents to support or substantiate the auditor's findings.

WORKING PAPER FORMAT

The precise format of working papers is generally not as important as is their completeness and legibility. First, the working papers are usually handwritten and include photocopies of several documents selected by the auditor to help substantiate the audit's findings. Very often the audit program manager will suggest or even require the general format to be used by the company's environmental auditors during the facility audits so that a basic level of uniformity is achieved. The clarity of working papers can be further facilitated by using paper of uniform size and appearance. Often, legal size paper is used so that copies of various documents can be easily fastened and included in the papers.

If all members of the audit team use the same general working paper format (including using a common paper size, initialing and dating all pages in one part of the page, etc.), the individual team members are more likely to feel they are contributing to the whole. Moreover, using a distinctive paper to record working notes can help remind part-time auditors of the need for special care in preparing audit notes.

A balance must be struck, however, between suggesting a common format and allowing the auditors the latitude to record their understanding, observations, and findings in the form most comfortable to them. Within a common overall format, each auditor should develop his or her own system for recording audit activities and observations.

Examples of completed audit protocols and the working paper documentation are provided in Figures 12-4 through 12-7.

FIGURE 12-4. Example of a completed segment of a protocol.

Air Pollution Audit Protocol (*continued*)

	Auditor(s)/ Comments	Working Paper Reference (List Page Numbers)
5. Prepare a list of all active, pending, and inactive regulatory permits, operating certificates, and registrations for airborne emissions (during the period under review). Include notation of agency, facility, type of permit, effective date, expiration date.		
a. Confirm that all identified emission sources are covered by permits.	JGG	JGG 7-12
b. Review for completeness the applications for registration, certification, operation, and construction.		
c. Determine that applications are accurate and signed by appropriate company personnel.		
d. Check compliance with PSD non-attainment rules.		
6. Develop a schedule of capital expenditures made during the period under review, and confirm that appropriate permits were obtained for all situations in which they were required.	GBE	GBE 5-6

(*continued*)

FIGURE 12-4. *(Continued)*

Air Pollution Audit Protocol *(continued)*

	Auditor(s) Comments	Working Paper Reference (List Page Numbers)
7. Using the emission sources developed in Step 3, and the permit data developed in Step 5:		
a. Determine that all applicable standards, limitations, and compliance schedules are being met for regulated emissions from all sources.	See working papers.	JMC 10
b. Determine through inquiry, observation, and review of available documents that the plant is in compliance with applicable opacity standards.	No citations were found and continuous sheets for opacity monitors showed no excursions)	
c. Determine that emission and fuel monitoring and test methods and procedures followed are in accordance with applicable regulations.	Certified analyses on composition of fuel met requirements.	
d. Review supporting documentation noting that all instrumentation, analytical techniques, calculations, records, and sampling locations and methods are in accordance with state and local regulations.	See working papers	JMC 10

FIGURE 12-5. Example of working paper documentation.

Item

7.a) Of the sources identified in 3 and 5, all met applicable standard limitation or compliance standards with the exception of the following:

1. Source A-2: Recent process changes have resulted in emitting volatile organic compound B instead of the permitted volatile organic compound A. Since the latter was given a limitation under NESHAP at about the same time as the proven change, no data have been obtained to establish compliance, nor has a revised permit application been submitted to the Air Pollution Control District (APCD) for County R.

2. Source B-4: Permit requires semi-annual submittal of test data on performance of Air Pollution Control System for particulate removal -- at the time of survey. These tests had not been carried out for the latest reporting period, nor was there any evidence that failure to meet the specified reporting had been communicated to the APCD for County R.

7.d) Selected charts from all automatic recording instruments required under permits (see Item 5) were reviewed and no evidence was found of out-of-compliance periods. A non-isokinetic sample for particulate sampling was used on all sample locations--however, the permit for Source D-3 requires isokinetic sampling. Although it is doubtful that there will be a significant change in results, especially since the measurements are well below limits, approval for this sampling method should be sought.

Records of instrument maintenance, calculations, and analytical techniques were inspected and found to be in accordance with all applicable regulations.

IMC-10

FIGURE 12-6. Example of a completed segment of a protocol.

Water Pollution Audit Protocol (*continued*)

	Auditor(s) Comments	Working Paper Reference (List Page Numbers)
b. Permits:		
(1) Develop a schedule of all permit requirements currently in effect;	*See working papers*	*IBH-4* *See Exhibits*
(2) Based on flow diagrams and observations, determine that operating procedures or installed systems are capable of providing information substantiating compliance with requirements;	*All systems shown on flow diagrams are capable of providing information*	*IBH-A, B, D*
(3) Ascertain status of compliance with all stipulations of permits;	*All stipulations are currently met*	
(4) For the review period, record date of all known excursions, type of excursion, abatement or corrective actions, and communications to regulatory agencies, etc. Determine if compliance with all reporting procedures has been carried out; if not, list exceptions/mitigating circumstances.	*See working papers*	*IBH-4*
(5) Obtain and review intra-company correspondence relative to permit limitation issues or regulatory actions, noting all unresolved issues.		*See Exhibits IBH-E, F, G, H*
A-2 Develop a flow chart or other description showing process and responsibilities for water pollution control activities (e.g., responsibility for sampling analysis, record maintenance, and regulatory reporting).	*SHG*	*SHG 4/5*
A-3 Based on information previously developed and your understanding of the system, confirm the operation of the data collection and reporting system:		
(a) Data collection:		
(1) With plant personnel, observe the procedure for sample collection, analysis, and data recording. Document the maintenance and calibration programs for the following devices: Composite sampling; Effluent flow measuring;	*IBH*	*IBH-4/5*

FIGURE 12-7. Example of working paper documentation.

Water Pollution Control Audit - Newtown Facility
Prepared by I. Ben Hadd
Protocol Step A.1 b (1)

Permit Schedules:

Discharge Point	Permit No.	Effective Date	Expiration Date	Parameters
001	CA17950-1	5/4/78	5/1/83	Flow, pH, TSS, BOD
002	CA17950-2	5/4/78	5/1/83	None (storm water damage)
003	CA17950-3	6/5/81	5/1/83	Flow (continuous) pH (continuous) BOD, TOC, TSS, Cu, Cr, Zn (24-hr composite)

Protocol Step A.1 b (4)

pH excursions were reported to regulatory agency (XXX) on 2/8/81, 5/6/81, 5/14/81, 7/20/81, and 8/5/81. All excursion periods reported were compared with recorder charts to determine accuracy of reported data. No discrepancies were noted.

In addition, a selected examination of pH recorder charts for 30 days (selected at random) during 1981 indicated compliance. All noted excursions were attributable to different process occurrences. Oral reports to regulatory agency (XXX) were recorded in internal memos; plant correspondence files indicate that all written reports were furnished within the established reporting period.

Concentration limitations for chromium were exceeded on 7/30/81. Delays in obtaining outside lab results resulted in reporting later than required by permit.

IBH-4

143

BASIC GUIDELINES FOR PREPARING WORKING PAPERS

Each audit team member is responsible for ensuring that the working papers generated during the audit provide sufficient documentation to substantiate the conclusions reached. In addition, each member should follow a few basic guidelines regarding the type of data gathered and the way in which it is recorded.

Keep Working Papers Clear and Understandable

Environmental audit working papers should be easy to read and understood by not only the auditor, but also anyone else reviewing the papers. Another person looking at the auditors' working papers should be able to tell, with little difficulty

The Scope of the Work Performed. What the auditor set out to do, which protocol steps the auditor was responsible for.

The Source of the Information. What the auditor did, which audit procedures were utilized, who provided information, where it was observed.

The Information and Facts. What specifically was found or observed, what was not obtained.

The Conclusions Reached. What the auditor concluded.

Providing this level of clarity requires a certain amount of discipline by the auditor to keep the working papers organized, neat, and legible. Taking good notes quickly and efficiently is a skill that should be cultivated. It is common practice among auditors to allow plenty of space as they are completing their working papers—by using only one side of a worksheet, writing on every other line, using only part of a page, or some combination of the above. Experience suggests that it is far better to leave parts of sheets blank in the beginning than to add something later on to an already full page.

Keep Working Papers Complete

Working papers should provide a complete picture of the work performed so that the conclusions reached are both reasonable and clearly substantiated by the notes and documents included in the working papers. They should be complete, free-standing records of what the auditor did. It is important to describe the systems fully rather than jumping to (or appearing to be jumping to) the conclusions. Working papers should document the good as well as the bad, and aid both the auditor and the reviewer in understanding the logic behind the conclusions.

Keep Working Papers Simple and Factual

It is critical that the auditor strive for simplicity and stick to the facts in preparing working papers so that an independent person reviewing the papers would reach the same conclusion the auditor reached. Jargon and unnecessary wording, listing, or scheduling should be avoided. The auditor should clearly identify the facts of the situation in brief sentences, diagrams, or tables. Working papers should be restricted to a factual description of those items that are directly related to the audit objectives. To illustrate what we mean by a "factual description," consider the following situation. You are standing in a valley looking out across a field at a house on the hillside and are asked what color the house is. Many would say something like "the house is white." The "correct" answer, based on information available, might be "white *on this side*." The basic message, of course, is to be able to relate the conclusion reached to the facts of the situation.

Significant care is required to keep the working papers simple and factual. Prudent language should be used that is consistent with the ultimate audit report purpose of clear and appropriate disclosure. For more information about prudent language, see Chapter 14.

Develop a Clear Context for Your Working Papers

In order to make working papers clearly understandable by another person, the auditor should develop a clear context for the papers. All exhibits should be identified and referenced in the proper context so that it is clear why the exhibit was selected and what it serves to help substantiate. In addition, the auditor should clearly cross-reference the working papers to the audit protocol or guideline so that it is easy to review these documents and determine whether all appropriate audit steps were completed, and find what segment of the working papers corresponds to a given protocol step.

In order to provide a clear and accurate context for the working papers, all documents and pages should be numbered and dated. The working papers should be arranged in such a way that each distinct audit step is addressed by a separate part of the working papers. As suggested above, the auditor should begin each step with a statement describing the purpose of the step, the scope of the work to be performed, the method(s) of gathering information, and the sources(s) of the information considered.

Prepare Working Papers While Doing the Field Work

Environmental auditors who have experienced the pressures of an audit may doubt their ability to consistently generate working papers such as those illustrated in Figures 12-5 and 12-7. All the general guidelines discussed above should be considered in the context of this important message: Write down your observations while performing your field work. Do not wait until the end of the audit—or even the end of the day—to document your understanding and findings. Experience suggests that understandable, complete, and factual working papers can be prepared as the audit is conducted. Keep in mind that the working papers are not a report that the auditor is to prepare from notes after the audit is completed. Rather, working papers are the auditor's field notes to keep track of audit procedures undertaken, results achieved, and items requiring further review or investigation. As such, it is important to develop and maintain the discipline to describe the audit procedures followed and record the facts and observations one step at a time. If the facility engineer is explaining a complex process faster than you can write it down, ask him or her to slow down. If necessary, take time immediately after the discussion to record your understanding. (Or even take a 15-minute break and come back later to confirm your understanding!)

DATA MANAGEMENT AND LEGAL CONSIDERATIONS

As implied throughout Chapter 12, the audit team will undoubtedly review an enormous volume of data (e.g., monitoring records, permits, reports, and maintenance records) within a relatively short time. In addition to having each audit team member take responsibility for preparing complete, accurate, and legible working papers, the audit team leader or program manager generally is responsible for reviewing all working papers for completeness, clarity, and consistency. During this review, the audit team leader should annotate the papers as appropriate (e.g., clarifications of ambiguous statements and cross-reference from one set of papers to another) and sign off as to their acceptability on each worksheet. Upon completion of this review, audit working papers are generally stored in a central file and kept under tight control by the audit program manager. Data management considerations are discussed further in Chapter 14.

In managing this large volume of data, it is important to recognize that several legal considerations can come into play. Working papers are an important record. They represent documentation of the scope and conduct of the audit, and if any questions arise at a later date, they may be needed to support the findings stated in the audit report. Because the working papers could be subject to subpoena or other disclosure outside the firm, it is essential that they be accurate, complete, and easy to understand. If the audit conclusions are challenged, factual, complete working papers will serve as a strong line of defense. Chapter 14 provides a discussion of confidentiality considerations and addresses specific legal protections that apply to working papers as well as audit reports.

CHAPTER 13

EVALUATING FINDINGS AND CONDUCTING EXIT INTERVIEWS

Two important activities in any environmental audit program are the evaluation of audit findings and the oral reporting to facility management at the conclusion of the audit. In the simplified audit scheme presented in Chapters 3 and 11, these activities are Steps 4 and 5 of the overall audit process. These are the two steps where the findings and observations of each member of the audit team are assimilated, reviewed, and reported.

Chapter 13 describes basic techniques and options for the auditor to consider in evaluating audit findings and presenting those findings to facility management. In addition, it presents examples of common roles played by the audit team and facility management during the exit interview.

EVALUATING AUDIT FINDINGS

Evaluating audit findings—Step 4 in the overall audit process—in our experience is often the one step that is shortcut, primarily because of time constraints. After the audit field work has been completed (Chapter 11), the auditors must wrestle with how best to organize the findings so that they are reported to appropriate levels of management (so that any deficiencies are corrected). During this step, the findings and observations of each auditor should be evaluated and their ultimate disposition determined.

As discussed in Chapter 7 (and, in particular, Table 7-1), the auditors should inform facility environmental staff of any deficiencies as those deficiencies are observed. Such communication is an integral part of a smooth, effective audit. Communication of potential problems and deficiencies should be an ongoing process between the auditors and facility staff. As was also discussed in Chapter 7, some items noted during the audit may require only the attention of the facility manager; others will require reporting to other levels of management and further follow-up. Preparing audit findings is an important step that requires significant time. It is best done when the auditor has a clear understanding of the reporting process that will be used.

After the field work portion of the audit is completed, each auditor should prepare his or her own list of findings. In preparing this list, it is crucial that the auditor ask the question, Have I gathered sufficient evidence to substantiate this finding? Each auditor should feel comfortable that, if challenged, he or she has the data to substantiate the finding.

Once these lists are prepared, the auditors should set time aside and together determine what they are going to say in the exit interview and how they will say it. A number of important questions now arise in determining how best to organize the findings. Is it really helpful to both report a departure from, for example, corporate policy, and then follow it up by citing 20 examples where the departure occurred? On the other hand, if only one to two examples are cited, will only these be corrected while the other 18 are overlooked? What messages will the audit findings convey about the facility staff? Will an audit resulting in six exceptions automatically be viewed as poorer facility performance than one with only three exceptions? If this is the case, perhaps the auditors should take extra care and attempt to group exceptions in comparable units. But will this grouping cause the problem to be buried in the exception and go unnoticed? Such are the issues the audit team must discuss and resolve.

Working from the lists of audit findings and observations of each auditor, look for situations where the sum may be greater than the parts. Several common findings (such as informal preventive maintenance systems for different types of equipment) when viewed as a group may have greater significance.

PREPARING FOR THE ORAL REPORT

Step 5 of the audit process introduced in Chapter 3 is to report audit findings. While the first activity is to apprise facility personnel during the audit, the first formal reporting generally occurs at the audit's exit interview with an oral report to facility management. The exit interview and oral report should be as carefully planned and prepared as any other step in the audit process. A handwritten summary of the audit findings, which may take a variety of forms, is often used to provide a framework for the oral report and to document what the auditors told facility management.

One of several available methods of summarizing and documenting audit findings is illustrated in Figure 13-1. This Exit Interview Discussion Form is prepared by the audit team leader and reviewed by each team member prior to the exit interview. Audit team findings are recorded on this form with working paper page references, and the disposition of each exception—either for the formal audit report or for local attention—is noted. As a working outline of the exceptions noted during the audit, this form usually is prepared in handwritten form at the end of the audit. This listing is often left with the facility as a record of the exit interview (such as Figure 13-1).

The audit team leader and team members should be prepared (and allow plenty of time) to discuss with facility management each exception noted on the exit interview form—the nature of the exception and the evidence found. The audit team should be prepared for facility personnel to comment on each exception, should take such comments under consideration, and should not leave a subject until questions have been answered and the exception has been clearly understood. When the overall purpose of the audit program calls for having the audit team provide recommendations for corrective action, the recommendations should be included as part of the exit interview and oral reporting session.

An alternative to the discussion form may be a summary outline of audit findings, as shown in Figure 13-2. Although such an outline is used within some companies, it does have some significant limitations. In particular, the audit team may be vulnerable since there is no documentation of the specific exception found by the audit team. Auditors do not want to find themselves in a position of having identified and explained (but not documented) a problem only to have the problem later arise as a complete surprise to facility management.

Some audit teams also prepare action plans on-site. These describe actions the facility should take to correct any deficiencies or to strengthen

FIGURE 13-1. Example of an exit interview discussion form.

	Exit Interview Discussion Form			Page _1_ of _3_

Facility: _XYZ Newtown Facility_   Audit Scope: _Air, water, Solid & haz Waste_

Audit Team Present: _John Smith, Jane Jones, William Day_

Facility Management Present: _James Allen, David Ross, Peter Mitchell_

Discussion Date: _March 23, 1983_   Prepared By: _John Smith_

W/P REF	EXCEPTIONS	FOR REPORT	FOR LOCAL ATT'N	COMMENTS/ DISCUSSION
JJ-15	1. the monthly fugitive dusting inspection of the bottom and fly ash basin are not being documented.	✓		We understand facility has recently established such a procedure and are in the process of implementing it
JJ-21	2. Recalibration of the in stack sulfur dioxide monitors are not being performed quarterly	✓		
WD-14	3. The system for reporting excursions beyond capacity limits is informal		✓	
WD-10	4. In touring the facility the audit team noted that transformers are not marked with an indication of PCB content	✓		

FIGURE 13-2. Exit interview outline.

Oral Report Outline

Facility: _X YZ Newtown Facility_

Audit Team: _J. Smith, J. Jones, W. Day_

Facility Management Present: _J. Allen, D. Ross, P. Mitchell_

Discussion Date: _March 23 1985_

I. Audit Scope

    1. _Air_

    2. _Water_

    3. _Solid and hazardous waste_

II. Audit Findings

    A. Deficiencies

        1. _Fugitive dust inspections_

        2. _Calibration_

        3. _SPCC Plan_

    B. Good Practices

        1. _Housekeeping_

        2. _Recordkeeping_

the environmental management systems. In these situations, action plans are developed by the audit team and facility personnel responsible for implementing the plans. Recommendations and action plans should be as specific, and actionable, as possible and identify those parties who should take the particular actions. Schedules and due dates should be set and procedures for carrying out the plans clearly spelled out. Figure 13-3

FIGURE 13-3. Example of an audit findings and recommendation form.

Audit Findings and Recommendations Form

Facility: _____

Audit Scope: _____

Audit Team: _____

Facility Management Present: _____

Discussion Date: _____   Prepared By: _____

Audit Finding:

_____

Recommendation:

_____

Corrective Action Plan:

   Effective Date of Corrective Action: _____

   Individual Responsible for Corrective Action:

   _____

_____

Comments:

_____

provides a means of recording audit findings and action plans. This form is then presented at the exit interview for discussion with and approval by facility management.

Depending on the audit objectives and scope, good practices of the facility may also be acknowledged during the exit interview and oral reporting session. If the formal written audit report does not include good practices, the exit interview discussion typically is limited to a general overall statement and description of the facility's good practices. If, on the other hand, good practices are included as part of the formal written audit report, the discussion tends to be more detailed and specific—and such comments may also be written on the Exit Interview Discussion Form.

It is important for the team leader to schedule an exit interview as far in advance as is feasible so that the facility manager and staff can plan their schedules accordingly. Once the audit team has completed its field work and can better predict the exact timing of the exit interview, the team leader should communicate such information to the facility manager.

THE EXIT INTERVIEW

The exit interview, or close-out discussion, is a meeting held at the conclusion of the audit among the audit team, facility manager, facility environmental coordinator, and other facility personnel who are key to the environmental activities at the location. In some companies, the audit program manager is also present. While some companies have chosen not to have written audit reports and use the exit interview as their only means of reporting, most use it as one step toward their formal audit reporting process (Chapter 14).

Exit interviews and oral reporting are important to the overall effectiveness and efficiency of the environmental audit. As such, it is important to provide sufficient time during the exit interview for a full discussion of all exceptions noted during the audit. The exit interview serves not only to communicate the audit team's observations to facility personnel, but also to foster relations between the audit team and facility personnel. Exactly what is discussed and how it is discussed varies from program to program, nevertheless, there are three basic parts to the exit interview: (1) setting the stage, (2) discussing the findings, and (3) describing next steps.

Setting the Stage

The exit interview should not be started abruptly. The audit team leader should work toward gaining the attention and rapport of the facility personnel by:

Acknowledging the cooperation obtained by the audit team from facility personnel.

Stating that all findings were discussed with facility personnel during the audit.

Portraying a willingness to discuss all issues in whatever detail is necessary.

Discussing the Findings

The exit interview is essentially an oral review of the findings and observations of the audit team. Copies listing summaries of the findings are often distributed to the participants at the exit interview. Each deficiency is discussed, including the reasons why it is a deficiency and the evidence found. If recommendations or action plans were developed, they are also presented. Each deficiency (and recommendation) should be fully and clearly communicated so that participants have a common view of the facts. Any comments made by facility personnel should be noted.

Describing Next Steps

At the end of the exit interview, the audit team leader should explain the overall reporting process and the facility's role in that process. This includes an explanation of how the report is prepared; what it will include; when it will be issued and to whom; and, if the facility is given an opportunity to comment on the report, the procedures and timing for such comments. Facility personnel should have a clear understanding of what will be included in the report before the report is issued.

The audit team should explicitly state and describe any next steps required—responding to the report, developing action plans, audit team follow-up, and so on. Facility personnel should be encouraged to consult with the audit team if any questions arise after the conclusion of the audit.

RESOLUTION OF CONFLICT

The exit interview is a particularly appropriate time for resolving conflicts and disagreements relating to audit findings. Occasionally, conditions at the facility may not have been interpreted correctly by the audit team, responses to questions may not have been fully understood, or a particular action (or inaction) may have a valid reason behind it. Often, such discrepancies are resolved well in advance of the exit interview. However, if not, the exit interview and oral reporting session provide an excellent means for the auditors to check with facility personnel their understanding of what they found. Interpretations and opinions of facility personnel should be welcomed and acknowledged. Such comments can be recorded on the Exit Interview Discussion Form so that the working papers reflect not only the auditors' observations presented at the close-out session but also points made by facility management.

Some audit programs prepare the written report and/or the action plan before the audit team leaves the facility. In such situations, conflicts need to be resolved during the exit interview. Conflicts will be minimized if the audit team is well-prepared for the exit interview and oral reporting session, the auditors' observations and findings are carefully thought out and developed, and criticisms of or "finger pointing" at specific individuals avoided. The exit interview should be viewed as a communications vehicle between the audit team and facility personnel—and not as a one-way or fault-finding lecture. The more involved the facility personnel become in the reporting session, the easier it will be for them to accept the findings and recommendations of the auditors.

Any exceptions or disagreements should be recorded. If differences of opinion or disagreements are unresolved at the conclusion of the exit interview, the audit program manager may have to take the matter higher up within the organization for resolution.

THE ROLE OF THE AUDIT TEAM AND FACILITY MANAGER

As previously discussed, the principal role of the audit team during the exit interview is to ensure that facility management has a clear understanding of audit findings and to try to resolve any misunderstanding or conflicts. An equally important role for the audit team is to engage facility personnel in the discussion of the audit deficiencies so that they participate and feel that their contributions are welcome. Insignificant findings should be downplayed and the cause of the underlying problem—not just the symptom—should be sought. In this search, facility personnel will have a good idea of what is going on, and the audit team should be attentive to those ideas.

The primary role of the facility manager is to understand the audit findings so that appropriate action can be taken by the facility. The roles of the facility manager during the exit interview may vary depending on the nature of the exit interview and the overall reporting process. Examples of some of the roles the facility manager may play are presented below.

Roles of the Facility Manager

Reporting Process	Facility Manager's Role
Report written off-site	Listens to the summary of the audit findings; questions and comments on the findings for further clarification and understanding
Report written on-site	Assists audit team in the preparation of the report; or reviews report with audit team during exit interview
Report and action plans written on-site	Prepares action plans; or signs off on the report and action plans

If the report and action plan are prepared on-site, the facility manager takes a more active role. Typically, the facility manager reviews the action plan to determine appropriateness, the time schedules for action, the individuals responsible, and the procedures for carrying the plan out. He or she may sign off on the action plan during the exit interview. This sign off documents that the facility manager knows of and agrees to the actions to be taken.

To the extent that your audit reporting process consists of an oral report only, the discussion presented above will assist in addressing issues to be considered for effectively implementing that process. If your program is structured so that the audit team prepares a written report, Chapter 13 provides an initial step to that process; Chapter 14 further addresses specific issues of the audit reporting process.

CHAPTER 14

AUDIT REPORTS

In Chapter 7, we described options for deciding to whom audit results should be reported and what information to include in an audit report. Here we describe how that information should be presented—how to provide clear and appropriate disclosure without unduly incurring additional risks or liabilities. Chapter 14, therefore, outlines some common reporting concerns, and some key principles to keep in mind when writing a report. In addition, it discusses report structure and a variety of special status reports to top management.

REPORTING CONCERNS

As discussed in Chapter 7, in general all audit reports seek to provide clear and appropriate disclosure of the audit findings by: (1) providing management information, (2) initiating corrective action, and (3) providing documentation of the audit and its findings.

Given these three foregoing objectives, the audit report must give due consideration to a number of concerns that are inherent in environmental audit reporting. Among those most frequently raised are whether certain disclosure requirements are being met, the risk of the "smoking gun," and the confidentiality of reports. When the report is written, these concerns must be addressed with sufficient clarity to ensure that undue liability is avoided.

Disclosure Requirements

Each facility has certain regulatory reporting requirements. Additionally, publicly held companies have certain SEC disclosure requirements (for example, the disclosure of any material effects of environmental regulation on the corporation, and all material environmental litigation). Moreover, the environmental statutes include provisions that require companies to maintain and provide the regulatory agencies information via permit applications and reports. (For example, NPDES permits require regular monitoring and reporting as well as special situational reports.) An initial concern for the auditor arises as to how the corporation will handle any external reporting that may be required of significant audit findings.

The Smoking Gun

An adverse report, indicating a facility does not comply with an environmental regulation, that has been buried in the files and not acted upon constitutes what is commonly known as the *smoking gun*. We believe this type of situation ought to be dealt with in a straightforward manner by both the environmental auditor and corporate management by making a prior commitment that a prompt review will be made and appropriate corrective action taken where noncompliance is observed and reported. The smoking gun risk will be minimized if an environmental auditing program is structured so that

  Corporate policy on the treatment of noncompliance items and other audit findings is clear.

  An audit report will automatically start the process for review of deficiencies and corrective action as appropriate.

  Top management is clearly apprised of the situation.

Confidentiality Concerns

In most situations, the environmental audit program has been established as a management tool for achieving compliance and, as such, we believe confidentiality concerns should not overly inhibit the conduct of the audit. It may be prudent, however, in selected situations or depending on specific circumstances of your program, to take certain steps to maximize protection of confidentiality. Potential protections include

Attorney-Client Privilege.   Which protects confidential communications between lawyer and client. In this case, the audit would have to be directed by and the audit report written for counsel. The protection of the privilege extends only to communications and does not prevent disclosure of facts. Use of this privilege, however, could limit communication of audit results to people who have a legitimate need to know.

Work-Product Doctrine.   Which provides qualified protection of documents prepared by or for counsel in anticipation of litigation. Note that the expectation of litigation must be present. Thus, the work-product doctrine is not a realistic protection for programs where audits are routinely conducted.

Self-Evaluative Privilege.   This emerging principle may give limited confidentiality privileges where the court believes the benefits of self-evaluation (audit) outweigh disclosure.

Physical Limitations on Distribution.   This potential protection, administrative in nature, aims to minimize chances of disclosure by limiting the actual distribution of the audit report and/or collecting all reports once the addressees have been apprised of their contents.

We offer another caution. Experts advise that there is no absolute way to ensure that an audit report will not become public at some future time, and all the potential protections listed above have limitations. Careful preparation rather than confidentiality devices is the best protection. Accordingly, to reduce the risks of audit reporting, it is important to describe carefully the exceptions so that they will be clearly communicated to and understood by the appropriate persons within the company who can take corrective action. In a recent survey of 13 companies who collectively conducted almost 600 audits in 1982, it was found that none uses confidentiality protections on a regular basis. Each makes an appropriate disclosure of the audit findings. And, each describes the specific exceptions found during the audit—rather than merely listing general topics—in its audit reports.

KEY PRINCIPLES IN AUDIT REPORT WRITING

Clear, Concise, and Accurate Reporting of Facts

The first principle is to report facts clearly and concisely. Every statement should be based on sound evidence developed or reviewed during the audit. Whatever is said must be supported or supportable. Speculation should be avoided. Generalities and vague reporting will only confuse and mislead the reader. For example, a report stating that "some" or "a few" or "not all" can leave the reader confused about the significance of the finding. Specific quantities should be used, such as, "of the ten samples taken, two were found to be. . .," "three of the five drums observed. . . ," and so on. A concern can arise in the reporting of, for example, "two deficient labels were found. . . .," leaving the reader with the impression that *only two* were deficient. On the other hand, saying one of a sample of ten could infer that the auditor believes 90% of the total were all right. Statements should be qualified as needed and any unconfirmed data or information should be so identified. Limitations of the representativeness of audit samples should also be identified.

Ideas or sentences that do not amplify the central theme should be eliminated. The report should not identify individuals or highlight the mistakes of individuals.

Certain reporting terms are especially useful in keeping generalities and ambiguities to a minimum in an audit report. Typical of these are the following.

Examples of Appropriate Report Phrasing

Do Not Say	When You Mean
The plant does not have. . .	We were unable to confirm that. . . We were unable to determine that . . . The audit team was unable to verify. . . Plant personnel were unable to locate copies of. . .
I found such and such to be true. . .	We understand that. . . We were told that. . . It appears that. . .

The plant is in compliance. . .	On the basis of our review, we observed the plant to be in compliance. . . On the basis of our review, it appears that. . . On the basis of X samples taken, it was found that. . .

Conversely, the auditor should abstain from using such language as is depicted below.

Examples of Audit Reporting Language to Avoid

Alarming	Gross negligence
Appalling	Incompetent
Careless	Intentional
Criminal	Neglect
Dangerous	Perjured
Deliberately	Reckless
Deplorable	Serious problem
Dishonest	Terrible
Disorderly	Violation
Fraudulent	Willful misconduct

Additionally, various terms are used in auditing programs and such terms should have uniform meaning among the various people involved in the audit program. Common terms used in most auditing programs are listed below.

Basic Environmental Auditing Terms

Verify	To establish accuracy or reality
Confirm	To prove or establish actual facts or details
Review	To go over or examine critically or deliberately
Examine	To inspect closely; to inquire into carefully
Finding	The results of an examination or review
Exception	Not conforming to a general rule

To keep your writing concise and liven up your writing style, follow these guidelines.

Shorten Phrases to Words

Instead of	Write
Emergency situation	Emergency
The reader of the report	The reader
A variety of options	Various options
As a means of providing	To provide
On occasion	Occasionally

Get Rid of Deadwood

Instead of	Write
On a daily basis	Daily
A person who is an experienced environmental auditor	An experienced environmental auditor
The functions are routine in nature	The functions are routine
A fine in the amount of $200	A $200 fine
At the present time	At present
In the month of June	In June
In regard to past due reports, we noted that they increased by 6%	Past due reports increased 6%
The total exceeds more than 100%	The total exceeds 100%

Avoid Verbs Camouflaged as Nouns or Adjectives

The *collection* of data will be accomplished next month.	The data will be collected next month.
The new procedure will accomplish a *reduction* of error entries.	The new procedure will reduce error entries.
Because of the change, an *improved* level of awareness will be achieved.	Because of the change, the level of awareness will improve.

A frequent *reconciliation* of the manifest documents should be performed.	The manifest documents should be reconciled frequently.

Avoid Indirect Expressions Where Possible

Instead of	Write
There were many instances of poor judgment found.	Many instances of poor management were found.
There are occasions when equipment is left unattended.	Equipment is left unattended occasionally.
It is expected that the new procedure will be in effect next week.	The new procedure is expected to be in effect next week.
It appears that the plant is in compliance.	The plant appears to be in compliance.

Three other guidelines will also prove useful.

1. *Use Short, Familiar Words.*   Big words usually are less readily recognized and understood than short words. But keep in mind that the key rule is ease of recognition and understanding. Dunch, roup, and cere are short words but will mystify all but Webster. *Instantaneously* is a long word but understood by most everyone—and it cannot always be replaced by short words such as now or quickly. So use short words when you can, but above all use familiar words.

2. *Write Digestible Sentences.*   Writing experts advocate an average sentence length of 15 to 18 words. Longer sentences run the risk of feeding the reader more ideas than he or she can digest at one time. Furthermore, if too many long sentences are used, the reader will become exhausted mentally. But remember, 15 to 18 is an average. Sentence length will and should vary so that some will be shorter and some longer.

3. *Be Complete and Accurate.*   In addition to being clear and concise, audit reports, like any report, must be complete and accurate. Completeness is more than checking for a missing word or sentence or number. It means taking into consideration the informational needs of any potential readers other than the primary addressees. It means placing findings in proper perspective so that, for example, an isolated incident does not

appear to be a chronic problem. Lastly, completeness means defining specific terms that are likely to be unfamiliar to the reader.

Inaccuracy appears in many guises. Incorrect data, misspelled words, misplaced decimal points, or a transposed number are common examples. A subtler form of inaccuracy is inconsistency of information in different parts of a report, or between the text and a table or chart. The reader cannot always identify which of the conflicting data are correct.

Establish a Context for the Report

The auditor should provide enough background information in the report so that the reader clearly understands who conducted the audit and what the audit did or did not include. The purpose of the report as well as the purpose and scope of the audit should also be described in a manner that enables the reader to know why the report was written and who should take corrective action.

Report in a Timely Manner

The timing of audit reports is critical to the overall reporting process and must be carefully thought out. Since management expects deficiencies to be corrected promptly, the report must be communicated in a timely fashion.

In many cases, a written draft of the audit report is prepared within one to three weeks of the completion of the audit. This draft then goes through a review and a final report is prepared. The final report typically is issued within five weeks of the audit's completion. The audit program manager should clearly communicate to those involved in the audit program the need for timely reports and develop specific deadlines for receiving comments on the draft report and issuing the final report. These deadlines must be adhered to in order that both the audit program and the report maintain their integrity.

Figure 14-1 presents the typical timing of some key steps in the audit reporting process. The timing that is presented is based on our survey of the reporting practices of leading environmental audit programs. While this timing is typical for each of the specific steps listed, there are variations among companies. It is important, however, that a target timetable be developed for your reporting process and the reporting process be managed against that timetable so that audit reports are consistently made in a timely manner.

FIGURE 14-1. Key steps in the environmental audit reporting process.

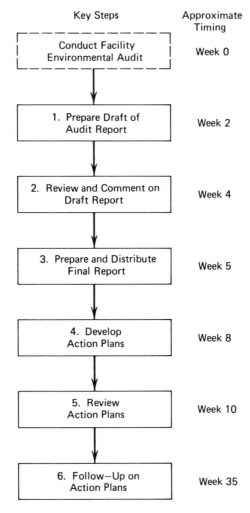

Key Steps	Approximate Timing
Conduct Facility Environmental Audit	Week 0
1. Prepare Draft of Audit Report	Week 2
2. Review and Comment on Draft Report	Week 4
3. Prepare and Distribute Final Report	Week 5
4. Develop Action Plans	Week 8
5. Review Action Plans	Week 10
6. Follow—Up on Action Plans	Week 35

Provide for Reviews and Follow-Up of Audit Reports

Audit findings must do more than paint an accurate picture of the environmental status of the facility. Any identified deficiencies must receive prompt consideration, action planning, and follow-up. Follow-up responsibilities also vary among companies as we discussed in Chapters 7 and 8.

Regardless of who is asked to review or reply to the audit report, the audit program manager should be explicit about the actions required of the addressees. Sometimes these actions are made explicit in the audit program design (e.g., "after receiving the report, a response is required within ten days"); other times the report itself states the specific actions required (e.g., "facility manager to submit a response by. . . ").

Audit follow-up procedures should be formalized and follow-up responsibilities clearly defined. Generally, a formal response to the audit report is made by the facility manager. In companies that require no formal response, action plans either are prepared on-site by the audit team and facility manager, or are incorporated directly into the final audit report as recommendations or specific action plans. In most instances, environmental affairs or line management assumes responsibility for ensuring that corrective action is taken—not the audit team.

Where the audit program has responsibility for monitoring follow-up, routine checks on action plans are typically initiated by the audit manager. Such checks can occur monthly, some within 60 days, and some every six months until the deficiencies are corrected. Occasionally, these checks result in a return visit to the facility by a representative of the audit program followed by a written status report on the actions taken. The status of action plans may also be incorporated in the report of a subsequent audit of the facility.

The direct involvement of the audit manager in action plans and follow-up varies among companies. In those companies where audits are repeated within a specified time, the auditor (or audit manager) commonly is directly involved in and central to action planning follow-up. However, if the audit scheme is such that an audit team is unlikely to return to the facility for some time, operating management typically assumes sole responsibility for follow-up.

Manage the Collection, Dissemination, and Retention of Audit Data

The audit program manager should establish policies on how data collected during the audit as well as the audit reports are to be managed. Documents should be stored and disposed of according to an established procedure. It may be advisable to have a formal records retention policy, so that records are kept as long as they are needed, and the records volume is manageable. (Records retention is discussed further in Chapter 19 of this book.) Regardless of the procedure used

All documents should be dated.

Preliminary drafts should be retained only until the material has been finalized.

Working paper documents should be stored in a central file.

REPORT STRUCTURE

The content of the report, described in Chapter 7, divides naturally into four basic sections. Each of these sections is described below and examples are provided.

Background

The first element of the report is often the background. Within the background section are the objective of the *report* and the objective of the *audit*, the scope of the audit, the conduct of the audit (where it was conducted, when it was conducted, and who conducted it), and finally the method employed (for instance, a physical survey of the plant, examination of records, interviews conducted, and any analytical tests performed). For example,

> This report summarizes the results of an environmental audit of XYZ Corporation's plant at Anytown, Massachusetts. Its objective is to communicate to XYZ Corporation management (1) the compliance status of the Anytown plant with respect to federal and state environmental laws and regulations, (2) the compliance status with respect to the corporation's environmental control policies and procedures, and (3) observations regarding environmental management systems. All deficiencies noted during the audit were discussed fully with facility management.

> The objective of the audit was to determine the facility's compliance with federal and state environmental regulations and corporate environmental policies. The audit included those air pollution control, water pollution control, and solid and hazardous waste management activities related to the Anytown plant during the period of January 1, 1983 through March 25, 1984.

> The audit was conducted on March 22 through March 25, 1984, and was based on: a physical survey of the facility; an examination of environmental, administrative, and operating records; discussions with facility management and staff; and verification procedures designed to examine the Anytown plant's application of and adherence to environmental statutes, policies, and standards.

Audit Findings: Environmental Statutes

The second key element of the report is typically the findings of the audit pertaining to environmental statutes. This is the auditor's opinion of the facility's compliance status and a list of exceptions. The exceptions cited in the report often include both an interpretation of the environmental statute to assist the reader in understanding what is required and a statement of the facts of the exception or the evidence found. For example,

> In reviewing a random sample of pH monitor records for 15 days, we identified three excursions of effluent pH that exceeded permit limits. We were not able to confirm that these incidents had been properly reported.

Audit Findings: Company Policies and Procedures

The third key element of the audit report is frequently the findings regarding compliance with company policies and procedures. If included in the scope of the audit, this section would include findings regarding adherence to company guidelines or conformance with internal standards (such as standards issued by the corporate engineering department). As with the previous section on regulatory requirements, this section includes findings, a list of exceptions, and a statement of the evidence found. For example,

> Corporate procedure requires the recycling of used oils. This procedure is not fully adhered to. The audit team observed four drums of waste oil mixed with refuse in the facility's disposal area.

Environmental Management and Control System

The final key element of the audit report is an evaluation of the environmental management and control system. This section includes observations and recommendations. The facts of the exception should be clearly defined. One way of correcting the deficiencies may be suggested. For example,

> Testing for the determination of wastes hazardous under RCRA appears to have been performed to meet the needs of the disposal firm, rather than to confirm or refute the basis for hazard determinations. We recommend that future testing be planned and undertaken with a clear statement of the facility's as well as the disposer's information needs.

Figure 14-2 provides an example of an audit report, combining each of the four key elements just described.

FIGURE 14-2. Example of an environmental audit report.

TO: The President
           XYZ Corporation

cc: James Allen, Manager
           Anytown Manufacturing Facility

SUBJECT: Environmental Audit of XYZ Corporation's Manufacturing
           Facility, Anytown, Massachusetts

DATE: May 7, 1984

---

1. Background

This report summarizes the results of an environmental audit of XYZ's manufacturing facility at Anytown, Massachusetts. Its objective is to communicate to XYZ's management: (1) the compliance status of the Anytown facility with respect to federal and state environmental laws and regulations; (2) the compliance status with respect to the company's environmental control policy and procedures; and (3) observations regarding the environmental management systems. All deficiencies noted during the audit were discussed fully with facility management.

The objective of the audit was to determine the facility's compliance with federal and state environmental regulations and corporate environmental policies. The audit included those air pollution control, water pollution control, and solid and hazardous waste management activities related to the Anytown manufacturing facility. Particular emphasis was placed on compliance with the Clean Water Act (CWA), the Clean Air Act (CAA), the Resource Conservation and Recovery Act (RCRA), and the Comprehensive Environmental Response, Compensation and Liability Act (CERCLA). The period under review was January 1, 1983 through March 23, 1984.

The audit was conducted from March 20 through 23, 1984. The audit team was composed of John Smith (Senior Environmental Auditor), Jane Jones (Auditor), and William Day (Auditor and Chemical Engineer), under the direction of John Smith. Participating staff of the Anytown facility included James Allen (Facility Manager) and David Ross (Facility Environmental Manager).

The audit was based on

A physical survey of the facility.

An examination of environmental, administrative, and operating records made available to us by XYZ staff.

Discussions with Anytown facility management and staff.

Verification procedures designed to examine the Anytown facility's application of and adherence to environmental statutes, policies, and standards.

FIGURE 14-2. *(Continued)*.

It must be noted that, at present, there are relatively few standards about what is "adequate" environmental management and virtually no generally accepted principles, procedures, or standards for conducting an environmental audit. Therefore, considerable judgment on the part of the audit team was involved in conducting the Anytown environmental audit.

II.   Audit Findings: Environmental Statutes

On the basis of our review, we believe that during the period January 1, 1983 through March 23, 1984, those environmental programs at the Anytown facility were such that the facility was in substantive compliance with existing federal and state air pollution control, water pollution control, and solid and hazardous waste management regulations or compliance schedules, as currently understood, except as noted below.

A.   Exceptions—Water Pollution Control

1.   In reviewing a random sample of pH monitor records for 15 days, we identified three excursions of effluent pH that exceeded permit limits. We were not able to confirm that these incidents had been properly reported.

2.   The SPCC plan states that the entire facility is fully fenced. In fact, the audit team observed that the no. 2 pond is not fenced.

B.   Exceptions—Air Pollution Control

1.   Three of several air emissions sources that were registered with the Department of Environmental Quality Engineering as controlled were not connected to control equipment.

2.   The Anytown facility is located in the Priority 1 area of Massachusetts Air Region I and is, therefore, required by Massachusetts Department of Environmental Quality Engineering regulations to have a written air pollution emergency plan of action available for inspection upon request. The written plan is to specify abatement strategies and courses of action for declared air emergency alerts, warnings, and emergencies. The facility was unable to produce a copy of such an air pollution emergency plan.

C.   Exceptions—Solid and Hazardous Waste Management

1.   In touring the facility, the audit team noted that transformers 1 and 2 outside of Unit A are not marked with an indication of PCB content. We were told by plant personnel that these are PCB transformers.

2.   A written waste analysis plan is required [40 CFR 265.13(b) (1-4)] for all hazardous wastes treated on-site or stored on-site by the Anytown

FIGURE 14-2. *(Continued)*

facility. The facility's current plan does not, in our opinion, meet the following requirements of 40CFR 265.13:

Rationale for selection of parameters to be analyzed.

Frequency of review and repeat analysis.

III.   Audit Findings: Corporate Policies and Procedures

On the basis of our review, we believe that during the period January 1, 1983 through March 23, 1984, environmental management activities at the Anytown facility were in conformance with Environmental Control Policy of XYZ Corporation, except as follows.

A.   Exceptions—Corporate Environmental Control Policy

1.   The corporate procedure for the recycling of used oils is not adhered to fully. The audit team observed four drums of waste oil mixed with refuse in the facility's disposal area.

IV.   Audit Observations: Environmental Management and Control Systems

The observations on the overall managerial approach and the Anytown facility environmental management and control systems cited below refer to matters that are not specifically required for compliance with one or more of the environmental statutes. However, they relate to areas of potential concern that, in the audit team's judgment, need attention or improvement.

A.   Observations and Recommendations—Solid and Hazardous Waste

1.   Testing for determination of wastes hazardous under RCRA appears to have been performed to meet the needs of the disposal firm, rather than to confirm or refute the basis for hazard determinations. We recommend that future testing be planned and undertaken with a clear statement of the facility's as well as the disposer's information needs.

B.   Observations and Recommendations—Environmental Management Systems

1.   Many of the managerial procedures and systems for environmental concerns pertaining to the Anytown facility are informal. Management systems tend to be based largely on the energies and attention to detail of a few key people rather than on written operating or environmental procedures. While this is often the case in facility-level environmental organizations and in no way is intended as a reflection on the capabilities of the Anytown personnel involved in the management and supervision of matters pertaining to the facility's environmental performance, the audit team recommends that

FIGURE 14-2. *(Continued)*

steps be initiated to formalize some of the components of the informal management and control system. Where standard practices are important to achieving compliance, the audit team believes they should be incorporated into operating manuals or developed into written standard operating procedures.

John Smith

John Smith
Environmental Auditor

OVERALL PROGRAM STATUS REPORTS TO SENIOR MANAGEMENT AND THE BOARD

While environmental audit findings and exceptions are reported primarily in individual audit reports, corporate officers and boards of directors have some special needs that can be addressed by overall program status reports. In this section, we will briefly discuss management's needs and the two general categories of audit program reports common to addressing those needs.

Needs of Corporate Executives

Corporate executives need to be assured that the environmental management functions are being conducted in accordance with applicable laws and regulations and corporate policies and guidelines, and that managers are fulfilling the corporation's environmental obligations. Corporate managers generally want to hear "good news" in environmental areas, be aware of "bad news," and know that problems are being addressed at the appropriate levels.

Board members are not often concerned, nor should they be, with the details of environmental matters. They are concerned with the potential effects on profits and losses to the corporation, with significant harm to the environment, and with the corporation's and their own personal legal liabilities.

They want answers to questions such as Is our environmental management program adequate? Does our environmental management program compare favorably to other companies' programs? What standards are used in making this judgment? (What is the basis for this judgment?) Will environmental matters have a material impact on the corporation? Are we in compliance with federal, state, and local laws and regulations?

Basic Types of Status Reports

Environmental audit program status reports to senior management and the board fall into two general categories (1) informing environmental management of what the audit program has accomplished and (2) advising management of significant findings.

FIGURE 14-3. Example of an activity report.

MEMORANDUM

TO: George Williams, Senior Vice-President

FROM: John Smith, Audit Program Manager

DATE: January 14, 1984

SUBJECT: Findings of Regulatory Deficiencies

In 1983, XYZ's audit program conducted audits at all of the corporation's 20 locations. The objective of the audits was to confirm that XYZ's environmental management system is in place throughout the corporation. The audit teams noted occasional departures from many applicable regulations as noted below.

Departures From	Number of Exceptions
CWA	(bar to 8)
CAA	(bar to 5)
RCRA	(bar to 10)
CERCLA	(bar to 3)

SUMMARY

CWA	8
CAA	5
RCRA	10
CERCLA	3
Total	26

The first category, *activity reports*, describes or summarizes the accomplishments of the environmental auditing program. These reports can be used to compare audits completed to audits scheduled, to illustrate the number and/or diversity of operations or activities audited, to describe audit coverage in terms of percentage of total facilities audited and/or audit cycle, to describe program budgets and other resources, or various combinations of these topics. Figure 14-3 provides an example of an activity report illustrating the number of regulatory deficiencies.

On the other hand, the second category, *evaluation reports*, is intended to provide an assessment (or input to an assessment) of the overall environmental performance of the organization. Current practice includes summaries and/or categorizations of audit findings and deficien-

FIGURE 14-4. Example of an evaluation report.

MEMORANDUM

TO: George Williams, Senior Vice-President

FROM: John Smith, Audit Program Manager

DATE: January 14, 1984

SUBJECT: Analysis of Audit Findings—1983

The following is an analysis of the findings from the environmental audits conducted in 1983:

		Deficiencies			
Function	Number of Sites	Regulatory	Procedural	Total	Average/ Audit
Air pollution control	6	5	5	10	1.7
Water pollution control	8	2	4	6	.75
Solid and hazardous waste	4	8	5	13	3.25

The functional area with the greatest number of findings per audit is solid and hazardous waste. Of the 13 findings, eight (62%) were noted at XYZ's Newtown Manufacturing Facility.

cies, analyses or differences in findings and exceptions among units or types of operations, and trend analyses. Figure 14-4 provides an example of an evaluation report summarizing audit findings by functional discipline.

The precise content, format, frequency, and mix between activity and evaluation reports appropriate for an environmental auditing program depends largely on the needs and expectations of senior management. Keep in mind that not all environmental audit programs include activity or evaluation reports to management. However, it is also important to remember that such reports can be useful in helping to meet the information needs of management and the board, as well as in initiating the management feedback and discussion that are necessary to ensure that audit efforts are congruent with senior management needs.

PART FOUR

ESTABLISHING AND ADMINISTERING YOUR AUDIT PROGRAM

CHAPTER 15

DEVELOPING A STRATEGY FOR ACHIEVING YOUR GOALS AND OBJECTIVES

A successful environmental auditing program depends on two factors. First, as discussed in Chapter 5, it depends on selecting program goals and objectives that are appropriate to your organization. Once the program objectives are defined, the second key success factor is to select from among the various design variables and alternatives discussed throughout this book those that most effectively support your program goals and objectives.

Devising the optimum strategy for your firm's audit program is a complex task. Environmental auditing is a relatively new endeavor; generally accepted or established principles, conventions, or standards have not yet been developed. As a result, there simply is not a universal strategy that works for all companies. In fact there are currently more questions than answers about why to audit, what to include in an audit, and how to audit. Moreover, in many companies, there will be different and sometimes conflicting views about the overall direction the audit program should take. As a result, special care is required to establish an effective audit program that achieves the objectives you and your company desire.

Chapter 15 presents a seven-step approach to developing a strategy for achieving your program goals and objectives. As in Chapters 16 through 20 it combines a narrative text with worksheets. Maximum value is gained by explicitly developing and recording your responses to the worksheets in this and subsequent chapters of Part 4.

IDENTIFYING MANAGEMENT'S EXPECTATIONS FOR THE AUDIT PROGRAM

The views and concerns of management that led to establishing an environmental audit program (or to approving your recommendation to establish a program) are important in shaping the audit program. We have found that it is crucial for the designer of a new audit program and the manager of an established audit program to understand clearly the specific role(s) senior management desires auditing to play. (*Note*: Our experience strongly suggests that an understanding of the role(s) that senior management desires environmental auditing to play is equally important whether the impetus for the program originated with senior management or the environmental staff. In either case, the audit program must function in a manner consistent with management expectations if it is to receive the support necessary for program survival and success).

Start your strategy development by first identifying the members of management who have been involved in the decision to establish a program. Then consider the primary concerns of management as they relate to environmental management in general and environmental auditing in particular. Take the time to develop a fairly exhaustive list. (Worksheet 15-1 can provide a useful framework.) In many companies a variety of individuals, many with different concerns—or at least different ways of formulating the concern—have participated in or concurred with the decision to conduct environmental audits. Specify the concern(s) of each individual listed in column 1, step 1, of Worksheet 15-1. If their involvement in the decision to audit is unknown, speculate about their concerns, and find ways to confirm your thoughts.

Next, for each management concern listed, specify the role that management desires auditing to play in addressing those concerns. (This is particularly important in companies where some of the initiative for establishing a program is coming from top management.) Be clear about the role(s) that auditing is to play vis-à-vis other environmental management programs and activities. For each role identified, critically assess

WORKSHEET 15-1. MANAGEMENT EXPECTATIONS

1. Key decision makers for my audit program include: (list all)

Function	Name(s)	Involve in Decision to Establish Audit Program			Comments
		Yes	No	Don't Know	
General Counsel	————	——	——	——	————
Environmental Counsel	————	——	——	——	————
Operations	————	——	——	——	————
Environmental Management	————	——	——	——	————
CEO	————	——	——	——	————
Board	————	——	——	——	————
Internal Auditing	————	——	——	——	————
Finance	————	——	——	——	————
Administration	————	——	——	——	————
Other	————	——	——	——	————
	————	——	——	——	————
	————	——	——	——	————

2. Primary concerns of my company's management that influenced the decision to establish an audit program are ————————————————
———————————————————————————————————————
———————————————————————————————————————

3. The specific role(s) auditing is to play in addressing those concerns is ———————————————————————————————————
———————————————————————————————————————
———————————————————————————————————————

4. What is it management wants the program to accomplish in the next 12 to 18 months? ———————————————————————————
———————————————————————————————————————

5. In management's view, to whom is the program ultimately responsible? ——————————————————————————————
———————————————————————————————————————

Summary: My management's expectations of the environmental audit program are as follows
———————————————————————————————————————
———————————————————————————————————————
———————————————————————————————————————
———————————————————————————————————————

whether an audit program can realistically fulfill that role. If not, you are very likely to have defined the role inappropriately.

Next, consider what management wants the audit program to accomplish. For example, is management more interested in finding problems so they can be remedied, or in providing assurance that the company's environmental management systems are working?

Keep your list of management expectations reasonably short. Ask, What is the minimum acceptable output of the audit program? What is it that management wants the program to accomplish in the next 12 to 18 months? Be clear about the priority or primary desires and record them in a brief statement on item 4 of the worksheet.

Now answer the question In management's view to whom is the program ultimately responsible? (item 5). For clues, ask, Responsible how or in what way(s)? In other words, How would the responsibility likely be managed? Record your understanding on the worksheet.

Finally, review your responses to items 1 to 5, and prepare a summary statement of your understanding of management's expectations of your environmental audit program. Ask how you can best confirm any assumptions that you have made.

UNDERSTANDING THE ORGANIZATIONAL CONTEXT

Begin this step by asking What is and isn't an acceptable role for your program? (See Worksheet 15-2.) For example, Can it function as a watchdog, policeman, trainer or educator, troubleshooter, detective, or pioneer?

For clues to acceptable roles, select several staff functions, other than yours (e.g., internal auditing and legal), and describe the basic role of each. Record your answers on the worksheet.

Next, examine management's philosophy regarding efforts that go beyond required compliance. Is this expected, encouraged, or acceptable? Acceptable as long as it doesn't cost too much? This philosophy can be particularly important given the "voluntary" nature of an environmental audit program. Develop your thoughts and record them under item 3 of the worksheet.

Now, critically assess how the performance of your program is likely to be measured. What will constitute a clear success? What is likely to be viewed as the line between success and failure? Go to item 4 on the worksheet.

WORKSHEET 15-2.   UNDERSTANDING THE CONTEXT IN WHICH YOUR PROGRAM WILL OPERATE

1. What is an acceptable or unacceptable role (e.g., policeman, watchdog, trainer or educator, troubleshooter, detective, or pioneer)?

    _____

    _____

    _____

2. What is the role of other analogous staff units (e.g., internal auditing, legal, security, risk management, or public affairs)?

    _____

    _____

    _____

3. What is management's philosophy regarding efforts that go beyond required compliance? _____

    _____

    _____

    _____

4. How is program performance likely to be assessed or measured?

    _____

    _____

    _____

Summary: The context in which the audit program will operate is

_____

_____

_____

_____

Finally, on the bottom of the worksheet, develop a summary statement that describes the context in which your program will operate.

DEVELOPING EXPLICIT PROGRAM GOALS AND OBJECTIVES

Table 15-1 elaborates upon range and variety of possible audit goals and objectives that appeared as Table 4-1. Review the candidate goals presented and sort them on Worksheet 15-3 as essential, desired, optional, or not desired. As you do this, remember that the goals and objectives

TABLE 15-1.   RANGE AND VARIETY OF AUDIT PROGRAM GOALS

Identification and documentation of compliance status, including:
   compliance discrepancies;
   differences or shortcomings at individual facilities; and
   patterns of deficiencies that may emerge over time.

Improvement in overall environmental performance at the operating facilities as a result of:
   providing an incentive not to allow problems to happen again
   reducing or containing problems that can interfere with normal business activity; and
   providing an incentive to clean up or improve housekeeping before an upcoming audit.

Assistance to facility management in:
   understanding and interpreting regulatory requirements, company policies and guidelines, and (perhaps) good practices;
   identifying compliance problems;
   defining cost-effective measures that should be taken to achieve compliance; and
   putting potential problems before a "committee of experts."

An increase in the overall level of environmental awareness as a result of:
   demonstrating top management commitment to environmental compliance;
   increasing the environmental awareness at the facility;
   the training accrued to the audit team; and
   involving employees in environmental, health, and safety issues.

Acceleration of the overall development of environmental management and control systems as a result of:
   auditing those systems that are "auditable,"
   defining the status of those activities that are not yet in a position to be audited;
   identifying important lessons learned and modifying systems and/or sharing information as appropriate;
   encouraging formulation of more formal procedures and standards for measuring environmental performance; and
   developing a data base of information on environmental performance that can be used in other mangement functions.

Improvement of the environmental risk management system as a result of:
   identifying conditions that may have an adverse impact on the company;
   assessing the risks associated with the hazardous conditions identified; and
   determining what actions are necessary to control those risks.

Corporate protection from potential liabilities as a result of:
   being able to demonstrate due diligence or evidence of the corporate envirmental commitment;
   soliciting an independent (third-party) opinion;
   documenting evidence demonstrating that the company is complying with regulations; and
   developing improved relations with regulators.

Development of a basis for optimizing resources as a result of:
   identifying current and anticipated costs and recommending ways to reduce those costs;
   identifying potential longer-term savings that can be accrued; and
   identifying potential opportunities to reduce waste generation.

WORKSHEET 15-3.   DEVELOPING EXPLICIT GOALS AND OBJECTIVES

A.   Sorting Through Candidate Goals

Potential Program Objective[a]	Essential	Desired	Optional	Not Desired
Identify and document compliance status	_____	_____	_____	_____
Improve facility performance	_____	_____	_____	_____
Assist facility management	_____	_____	_____	_____
Increase awareness	_____	_____	_____	_____
Accelerate the development of management and control systems	_____	_____	_____	_____
Improve risk management system	_____	_____	_____	_____
Enhance protection from potential liabilities	_____	_____	_____	_____
Develop a basis for optimizing resources	_____	_____	_____	_____
Other _____	_____	_____	_____	_____
Other _____	_____	_____	_____	_____

B.   Developing explicit program goals and objectives

1.   Long-term goals _____

_____

_____

2.   Program output for the next 12 to 18 months _____

_____

_____

3.   The output of an individual audit _____

_____

_____

_____

My overall program goals, expressed as concretely as possible and in operational terms, are as follows

_____

_____

_____

_____

[a]Refer to Table 15-1.

should be critical, specific and measurable, controllable to the person responsible, and directed toward a specific end result.

Then, starting with the basic goals you have identified as essential and the guidelines for goal development presented above, revise and refine the wording of the goals. Be as concrete and operational as you know how to be. (To help sharpen your wording, first ask whether you would accept those goals from a subordinate and then whether you would be willing to have your performance review depend on their achievement.) Refer to Worksheet 15-3 and sort out long-term goals; define the output of the program over the near term (next 12 to 18 months). Do the differences between long-term and near-term goals seem reasonable? Define the output of an individual audit. What is it that an individual audit is to accomplish?

DEFINING LINKAGES AND OVERLAPPING RESPONSIBILITIES

With what organizational entities must your program effectively interact to be successful? To find out, you need to define the linkages and areas of overlapping responsibilities that will exist between your audit program and other management programs within the company.

Start by asking what sources of data will be important to the program. Will your program require data inputs for selecting sites or topics for in-depth auditing, for assessment of risks, and so on? Note your responses on Worksheet 15-4.

Next, make a cut at who will be responsible for action planning and follow-up on various exceptions found during an audit. In some companies, these roles are played primarily by people within the audit program, in others by individuals outside the audit program (e.g., a district or regional environmental engineer). In yet others the responsibilities are shared. Describe on the worksheet how you envision (1) audit action planning and (2) follow-up will be accomplished at your company.

Similarly, ask yourself who will disseminate "lessons learned" from one environmental audit to others (such as other facilities with similar environmental situations to an audited facility) and how they will be disseminated. Record your thoughts on item 3 of Worksheet 15-4.

Then consider the various groups that may be called on to provide audit team staff to the program. List the likely groups on Worksheet 15-4.

WORKSHEET 15-4.   DEFINING LINKAGES AND OVERLAPPING RESPONSIBILITIES WITH OTHER PROGRAMS

1.   Data sources _____

_____

2.   Action planning and follow-up on audit findings will be done by

_____

_____

3.   Lessons learned from environmental audits will be disseminated to others who should know by _____

_____

4.   Auditing team staffing will come from the following groups _____

_____

_____

5.   Differences of opinion about findings will be resolved by _____

_____

_____

In order to be successful, my audit program must effectively interact with these organizational entities

_____

_____

_____

Next, identify how and by whom differences of opinion (e.g., about audit findings, conclusions, or recommendations) are likely to be resolved. Imagine that a plant manager and the audit team leader disagree about the desirability or appropriateness of a recommendation. Who will decide? Will a third party (perhaps the audit program sponsor, the director of operations, or the company's senior environmental manager) be called on to resolve the difference?

Finally, list at the bottom of Worksheet 15-4 those organizational entities with which your audit program must interact in order to be successful.

IDENTIFYING POTENTIAL BARRIERS TO IMPLEMENTATION

What obstacles can be anticipated to implementing your audit program? To what extent will your program's implementation be viewed by others as a significant new program or major organizational change? What attitudes can be anticipated on the part of plant management, division management, corporate environmental staff, and other corporate staff? What are the two or three most significant obstacles for implementation

WORKSHEET 15-5.   IDENTIFYING THE POTENTIAL BARRIERS TO IMPLEMENTING YOUR PROGRAM

1.   Is this likely to be viewed as a major organizational change? (If yes, how does your company typically go about implementing change, e.g., negotiation, phased implementation, top-down.)

_____

_____

_____

2.   Facility management attitudes are likely to be _____

_____

_____

_____

3.   Division management attitudes _____

_____

_____

_____

4.   Corporate environmental staff attitudes _____

_____

_____

_____

5.   Other corporate staff attitudes _____

_____

_____

_____

Key obstacles that can be anticipated are

_____

_____

_____

that can be anticipated? What major objections to the program can be anticipated? Record your answers on Worksheet 15-5.

KNOWING THE LIMITATIONS AS WELL AS THE STRENGTHS OF YOUR PROGRAM

To identify the limitations as well as the strengths of your audit program, begin by asking to what extent environmental activities and programs are "auditable" within your company. Using Worksheet 15-6, make two lists. One list indicates the key aspects of your environmental compliance and risk management program that are basically auditable (systems, procedures, records, etc.), the other indicates areas likely to be not yet auditable.

WORKSHEET 15-6. KNOWING THE LIMITATIONS AS WELL AS THE STRENGTHS OF YOUR PROGRAM

1. (a) Auditable environmental programs and activities _____
    _____
    _____

    (b) Programs and activities not yet auditable _____
    _____
    _____

2. Role of verification _____
    _____
    _____
    _____

3. Status of corporate environmental policies and procedures _____
    _____
    _____
    _____

4. Facility's knowledge of regulatory requirements and corporate environmental policies and procedures _____
    _____
    _____

Key strengths and limitations of your program
_____
_____
_____
_____

Then examine how much verification you anticipate will be done as part of your audit program. Also indicate on the worksheet the status of the corporate environmental policies and procedures and the facility's knowledge of these requirements. Finally, on the bottom of Worksheet 15-6, summarize the key strengths and limitations of your program.

COMMUNICATING YOUR PROGRAM

A successful strategy includes "selling" your program up, down, and across the organization. To be an effective salesperson you must know both your product and audience.

Review your responses to Worksheets 15-1 through 15-6 and note on Worksheet 15-7 those persons it is important to "sell." Focus your communications on needs of and benefits to the corporation and various levels of management.

WORKSHEET 15-7.   PROGRAM COMMUNICATIONS

1. Vehicle(s) and responsible person for communications higher up within the organization _____

    _____

    _____

    _____

    _____

    _____

2. Vehicle(s) and responsible person for communications down the organization _____

    _____

    _____

    _____

    _____

3. Vehicle(s) and responsible person for communication to each critical linkage

    _____

    _____

    _____

    _____

    _____

CHAPTER 16

ESTABLISHING YOUR ENVIRONMENTAL AUDITING PROGRAM: GETTING STARTED

A successful environmental auditing program is built on a foundation of demonstrated support from management, clearly delineated responsibilities and accountabilities, and a strong technical framework. The worksheets in Chapter 15 were aimed at clearly defining the program objectives, management's expectations for the program, the organizational context within which the program functions—including roles, responsibilities, and linkages with other departments, and the barriers to implementation.

Chapter 16 begins where Chapter 15 left off, and outlines the steps to go through in developing the basic framework of your audit program and defining your short-term objectives and milestones. The basic elements of an environmental audit program have been described in Parts 2 and 3 of this book; thus Chapter 16 provides a guide for selecting elements from the earlier parts of the book.

AUDIT PROGRAM FRAMEWORK

Having gained a clear understanding of what management expects the program to accomplish and what the basic program responsibilities and organizational linkages are, the next step is to select from the many alternative program elements, those best suited to accomplishing your program objectives. Throughout this section, focus on the framework you want to have in place for the *next 12 months only*. It is better, and ultimately more productive, to start small and then add to your basic framework once the program is established. However, to the extent that you can identify program elements you eventually want to have in place (but not necessarily this year), note them as such.

To develop the basic framework, pay specific attention to three general areas: (1) audit boundaries, (2) staffing, and (3) audit procedures.

Audit Boundaries

Using Worksheet 16-1 as a guide, begin by reviewing Chapter 5 in order to define the scope of your audit program. What is it that needs to be included in the program scope during the next year? Will the program include the entire corporation or just one or two segments? Which functional disciplines need to be included this year, recognizing that others can be added at a later time? Which compliance parameters will the team audit against? Do you expect the audit team to look beyond the property line of the plant?

Next, review the section of Chapter 11 on "Data Gathering Methods." In order to fulfill the program objectives defined in Chapter 15, in what depth do the auditors need to go? What percentage of their time should be devoted to inquiry, to observation, and to testing? Finally, compare the resources you expect to be available to the audit program over the next 12 months with the resources required to fulfill the audit program objectives.

Staffing

The second basic program design element you need to focus on is staffing. Review Chapter 6 and use Worksheet 16-2 as a guide to selecting audit team leaders and members. In assessing the types of expertise you would like to have represented on the team, distinguish between what expertise each team member must have and what expertise at least one of the

WORKSHEET 16-1. AUDIT BOUNDARIES

1. Define the scope of your audit program.

   (a) *Organizational.* Entire corporation? If not, list which parts of the organization are covered.

   _____
   _____
   _____
   _____

   (b) *Functional.* Which disciplines are included?

	Yes	No
Air pollution control	___	___
Water pollution control	___	___
Solid and hazardous waste management	___	___
Occupational health	___	___
Occupational safety	___	___
Product safety	___	___
Other _____	___	___
Other _____	___	___
Other _____	___	___

   (c) *Compliance Hierarchy.* Which applicable compliance parameters will you audit against?

	Yes	No	N/A
Federal laws and regulations	___	___	___
State laws and regulations	___	___	___
Local requirements	___	___	___
Corporate policies and procedures	___	___	___
Division policies and procedures	___	___	___
Facility policies and procedures	___	___	___
Industry standards	___	___	___

   To the extent possible or reasonable, define which policies, procedures, or industry standards are applicable.

   _____
   _____
   _____

   (d) *Locational.* Do you expect the audit team to look physically beyond the property line of the plant?   Yes _____ No ___
   If yes, what specifically should the team look at?

	Yes	No
Other nearby property owned by the plant	___	___
Transportation routes	___	___

Hazardous waste disposal facilities   ___ ___
Nearby environmentally sensitive areas   ___ ___
Nearby residential areas and recreational facilities   ___ ___
Other nearby property leased by the plant   ___ ___
Distribution centers   ___ ___
Contract packaging or toll manufacturing activities   ___ ___
Other _____   ___ ___
Other _____   ___ ___

2. *Depth of Review.* Given the objective(s) of your audit program, estimate what percentage of the auditors' available time should be spent on
     Inquiry   _____
     Observation   _____
     Testing   _____
                100%

3. *Resources*
    (a)   Estimate the approximate number of person days required to audit each of the functional areas listed in step 1(b) above (recognizing that it will vary from site to site).

Functional Area	Person Days
_____	_____
_____	_____
_____	_____
_____	_____
_____	_____
_____	_____
_____	_____

    (b)   List the resources available to the audit program over the next 12 months.

Name of Person	Days Available
_____	_____
_____	_____
_____	_____
_____	_____
_____	_____
_____	_____
_____	_____

    (c)   What resources are required to fulfill program objectives?
         1. Total person days required to audit the functional areas [total of step 3(a)]   _____

2. Add time spent on-site doing general and administrative activities (e.g., meetings with plant management, team meeting, and plant tour) _____
3. Total person days on-site for an average audit [add (1) and (2)] _____
4. Total person days required for preparation and follow-up _____
5. Total person days per audit [add (3) and (4)] _____
6. Number of facilities to be audited in next 12 months _____
7. Total resources required [multiply (5) × (6)] _____

(d) Compare the total resources available from step 3(b) with total resources required to fulfill program objectives from step 3(c). Summarize any differences below.

_____

_____

(e) Estimate the funding required versus funding available for travel, lodging, and so on.

_____

_____

WORKSHEET 16-2. STAFFING

1. Which of the following areas of expertise are desired for the audit team leader (see Chapter 7)?

	Essential	Desired	Optional
Legal considerations	_____	_____	_____
Environmental control technologies	_____	_____	_____
Scientific disciplines	_____	_____	_____
Facility operations	_____	_____	_____
Auditing	_____	_____	_____
Management systems	_____	_____	_____
Knowledge of peer facilities	_____	_____	_____

2. Which areas of expertise are desired for team members?

	Essential	Desired	Optional
Legal considerations	_____	_____	_____
Environmental control technologies	_____	_____	_____
Scientific disciplines	_____	_____	_____

WORKSHEET 16-2. *(Continued)*

Facility operations	_____	_____	_____
Auditing	_____	_____	_____
Management systems	_____	_____	_____
Knowledge of peer facilities	_____	_____	_____

3. List the individuals in your company who are in a position to know of people who are potential team members and who have the authority to make people available to the program.

Individual	Knows of Potential Team Members	Can Make Them Available
_____	_____	_____
_____	_____	_____
_____	_____	_____
_____	_____	_____
_____	_____	_____
_____	_____	_____

4. List potential team leaders and team members.

Individual	Potential Team Leader	Potential Team Member
_____	_____	_____
_____	_____	_____
_____	_____	_____
_____	_____	_____
_____	_____	_____
_____	_____	_____

auditors must have. Distinguish the desired qualifications and expertise for the team leader from those of the team members. Determine who in your company may know of potential team members and who is in a position to make them available to the audit program. Finally, develop a list of individuals who are potential team leaders or team members.

Audit Procedures

Before you can go on the first audit, you need to develop the basic tools required to prepare for, conduct, and follow-up on the audit. These tools are the basic working documents and procedures used by the audit team leader and team members for each audit.

Using the checklist below as a general guide, carefully review Chapters 9 through 12 and develop the basic working documents needed to conduct an environmental audit. What documents will you need as you prepare for the first audit? What type of audit protocols or guides would you like to have? How can you tailor the protocols and questionnaires provided in the Appendixes to meet your needs? Determine how you expect the on-site activities to be managed. What type of training is required in order to make sure the audit team works effectively and efficiently on-site? What type of audit working papers do you expect from the auditors?

Audit Procedures Checklist

A. *Pre-Audit Activities* (Chapter 9)

_____   1.   Take Table 9-1 (example pre-audit reminder list) and tailor it to your specific audit program needs.

_____   2.   Review each of the items on the list looking for ones that could cause potential problems.

_____   3.   Develop necessary back-up worksheets.

_____   (a)   List of information to request in advance.

_____   (b)   List of items to take to the plant.

_____   (c)   Other _____

B. *Audit Protocols and Questionnaires* (Chapter 10)

_____   1.   Carefully review the alternative audit protocol formats and styles listed in Table 10-1 and described in Chapter 10.

_____   2.   Identify which format(s) you feel are best suited to your needs and which are most readily available. (*Note*: Copies of the basic protocol and the yes-or-no questionnaire are provided in the Appendixes.)

_____   3.   Develop the audit guides (or modify as appropriate those provided in the Appendixes) so that you are ready for the first audit.

C. *Field Work* (Chapter 11)

_____   1.   Carefully review the general approach to field work and the data gathering methods and techniques described in Chapter 11.

(continued)

_____ 2. Determine the overall approach to field work for your audit program including what things to emphasize and what must be consistent from audit to audit.

_____ 3. Determine the nature and level of training required to bring the audit team leader(s) and members "up to speed" regarding field work.

D. *Working Papers and Audit Record Keeping* (Chapter 12)

_____ 1. Review the working paper descriptions and examples provided in Chapter 12 and decide what level of detail the working papers need to be to support your program objectives.

_____ 2. Determine the type of working papers desired (e.g., notes kept directly on the audit guide or protocol, comments kept on separate notepads of uniform type).

_____ 3. Determine what type of supervisory review and/or quality control (if any) is needed or desired to ensure that the audit working papers are complete and appropriate.

CHAPTER 17

SELECTING AND SCHEDULING INITIAL FACILITIES FOR AUDIT

Once your audit program has been designed, supporting materials developed, and an organization and staffing approach selected, you are ready to reduce the program concept to practice. A number of issues should be considered as you select and schedule the initial one or two audits. Chapter 17 describes these issues and provides worksheets to help guide audit program designers and managers through the process.

SPECIAL CONSIDERATIONS FOR THE FIRST AUDIT

The first audit in a new environmental auditing program can be especially important because it often involves special objectives. Worksheet 17-1 provides a useful framework for considering the special outcomes that should be associated with your initial audit.

To identify the special objectives or outcomes that are important for your company's first environmental audit review the basic objectives for your audit program. Consider the reasons for establishing your audit

WORKSHEET 17-1. OBJECTIVES FOR THE PILOT AUDIT

1. What are the basic objectives that have been established for your program? (Review your response to Worksheet 15-1.)

_____

_____

_____

_____

2. The expectations of key individuals involved in establishing the audit program are as follows:

Person	Expectations for the First Audit
_____	_____
	_____
	_____
_____	_____
	_____
	_____
_____	_____
	_____
	_____

3. To what extent are the following outcomes important to accomplish during the first audit?

	Very Important	Somewhat Important	Not Required
Train the team in auditing	____	____	____
Familiarize the team with environmental considerations	____	____	____

WORKSHEET 17-1. *(Continued)*

	Very Important	Somewhat Important	Not Required
Debug the audit protocols and questionnaires	____	____	____
Test or validate the audit approach	____	____	____
Conduct a complete and thorough audit	____	____	____
Find problems and concerns	____	____	____
Develop examples for subsequent training	____	____	____
Other	____	____	____

program. Identify those individuals who were influential in making the decision to start an audit program as well as those who helped design the program. Ask yourself whether it is acceptable for your company's first environmental audit to be positioned as a "pilot" audit or learning experience—or whether the team must complete the entire audit process. That is, is it acceptable for the team to learn a lot (either about the facility's environmental status, or about auditing, or perhaps both) but not complete the audit?

Next, consider the basic audit approach that you have developed and the staffing alternative you have selected. What needs, if any, are there to validate the basic approach that you have chosen? To what extent are audit protocols, questionnaires, guides, and other audit materials likely to require debugging? How familiar will the members of your first audit team be with the audit process and the various materials that have been developed to guide them in conducting the audit?

Experience suggests that the learning curve for environmental auditing is relatively steep, and that significant learning occurs during the first several audits. For this reason, it is important that your objectives for the initial audit(s) be achievable.

SELECTING THE FIRST FACILITY TO BE AUDITED

In general, it is desirable to select as the initial facility to be audited one that is representative of company operations, environmental complexity, and local management awareness of environmental matters—in other words, a "typical" facility. Worksheet 17-2 provides a framework for

WORKSHEET 17-2. FACTORS FOR SELECTING FACILITIES

Factors	Relative Importance[a]	Facility			
Size of facility					
Complexity of environmental considerations					
Air pollution relative risk (inherent)					
Water pollution relative risk (inherent)					
Solid and hazardous waste relative risk (inherent)					
Occupational safety relative risk (nature of manufacturing hazards)					
Occupational health exposures to toxic chemicals and noise					
Product safety realtive risk					
Size of facility environmental, health, and safety staff					
Attitude and support of facility management					
Public sensitivity to the company and to environmental issues					
Other					
Other					

[a]High, medium, low.

considering and evaluating factors that help you select such a typical facility. Consider which factors are most important. Is it more important that the facility is of average size or that it represent an average environmental risk or that management have a good attitude toward the audit?

The first audit can be difficult and challenging. Unless you are using outside assistance to develop your program or have hired staff from a company with an audit program, chances are no one will have conducted an in-depth environmental audit before. Despite the amount of development training conducted in advance of the review, the audit team is likely to be somewhat unfamiliar with the protocols and procedures.

TABLE 17-1.   ELEMENTS OF KEY SUPPORTING DOCUMENTS FOR SCHEDULING THE AUDIT

Letter to Facility Manager Confirming the Audit

Date and starting time of the audit
Team leader and team members
Purpose of the audit
Audit scope
Statement of why the particular facility was selected
Overview of audit process
Notation of times where interaction with the facility manager would be especially important (e.g., opening conference, daily wraps, and exit interview)

Confirmation Letter to Audit Team Members

Location, dates, and time of audit
Audit purpose
Overall team staffing and assignments
Discussion of information and advance materials, if any, to be collected
Preparation required or desired
Schedule of audit planning process (noting dates where additional information will be provided and/or inputs will be required)

Opening Conference Presentation Materials

Purpose of audit
Scope of audit
Team leader and team members
Selection process and planned audit frequency
Brief history of why the audit program was established
Overview of basic audit process
Explanation of the audit reporting process

SCHEDULING THE INITIAL AUDIT

It is important to allow enough lead time for the first audit. As noted, the first audit involves special considerations beyond those general scheduling factors discussed in Chapter 9. Not only must all the initial arrangements be made (scheduling the audit, notifying facility management, notifying members of the audit team and confirming their availability, assembling and reviewing background information, etc.), but also, because you have not done them before, you will not be able to go to the files for a model to follow.

Paying attention to not only what you say but also how it is said in scheduling audits can be important in helping build the rapport necessary to the success of your program. Moreover, careful preparation of the initial communications can provide the audit program manager with models for various audit communications that will be repeated with each audit. Table 17-1 provides an overview of three basic supporting documents that most audit managers will need if they are to schedule the initial audit efficiently.

CHAPTER 18

DEVELOPING A TYPICAL SCHEDULE AND AUDIT CYCLE FOR YOUR PROGRAM

Parts 1 and 2 discuss a number of important aspects that affect the selection of an audit cycle and development of a typical schedule for your audit program. Chapter 18 pulls this material together and provides a guide for determining the audit cycle and typical schedule appropriate to your company.

AUDIT ALL FACILITIES?

There are two schools of thought on the issue of whether to audit all operating facilities on some sort of repeat cycle. Those who view the audit as an important tool for confirming that compliance is being achieved and appropriate environmental management systems are in place and functioning, favor regular reviews of all key aspects of air pollution control, water pollution control, hazardous waste management, occupational safety, occupational health, and product safety at each operating facility.

The other school of thought views the environmental audit more as a spot-check. Here operating management is expected to take the necessary steps to confirm regularly that appropriate management systems and procedures are in place and being followed. This school of thought favors reviews of selected aspects of selected facilities.

The starting place for sorting out where your program fits is your program objectives and the expectations of management. Using Worksheet 18-1, reexamine the goals and objectives that you have developed. Clarify or otherwise revise your objectives to eliminate ambiguities. "Assurance to management" may require only spot-checks, particularly if a basic set of environmental procedures and practices apply across several operating facilities. "Verification of compliance" may be carried out over a staggered schedule with permit performance reviewed regularly and permit requirements audited again only when either the permits or the operations are modified. On the other hand, phrases such as "no surprises" may require that audits be fairly frequent (or almost continuous, if the expression is taken too literally).

After sharpening your understanding of the outcome that is desired, consider the resources required versus those that are likely to be made available. Adjust your estimates to reflect the sources(s) of the audit staffing and any special requirements for training or preparation. Make a similar estimate of the dollar budget that you believe will be made available. Then, take a broad accounting of the staffing requirements for the audit including not only estimated time to be spent in the field by the audit team but also team preparation and follow-up. If your estimates of the required and available resources are relatively close, you will probably discover that some fine tuning of audit objectives and outcomes versus number of audits and resources required per audit will provide the necessary balance. If available resources differ significantly from resources required, major adjustments are needed.

Having come at the audit cycle from the perspective of the audit team and reached a balance of resource requirements and availability, it can be useful to shift perspectives. This time consider availability versus requirements from the perspective of the facilities that will be audited. Here again, delineate time required for audit preparation, the actual audit, and audit follow-up. In considering availability, factor in both the general staffing level of the facility's environmental management group and the nonenvironmental requirements that have been placed on others from the facility who are likely to interface with the audit team or otherwise play a significant role in the audit. Frequent audits may not make much sense for facilities with particularly lean staffing. On the other hand, the

WORKSHEET 18-1. DETERMING AN AUDIT CYCLE

1. What are your primary audit program objectives?

_____

_____

_____

2. Given these objectives, which of the following most accurately describe the role of your audit program?

	Yes	No
A spot-check on the system	___	___
A check to make sure the facility is basically doing the "right" things	___	___
An integral part of the compliance management system	___	___
A check to make sure there are no surprises	___	___

3. What resources can be made available?

(a) *People.*   Make an estimate of the person days available for
    Audit preparation   _____
    On-site audit   _____
    Post-audit activities   _____

(b) *Money.*   What dollar budget is likely to be made available? _____

4. What time (in person days) is required to conduct an average audit?

	Air	Water	Hazardous Waste	Other
(a) Audit preparation	_____	_____	_____	_____
(b) On-site audit	_____	_____	_____	_____
(c) Follow-up	_____	_____	_____	_____

5. What corporate or other audits are the facilities subjected to?

_____

_____

_____

leanness itself may be prompting a desire on the part of management for regular audits of all facilities.

Lastly, it is useful to consider the number of other (nonenvironmental) recurring "audits" that a facility undergoes. Here, it is useful to view the word in its broadest sense. The development of more than one environmental audit program has been temporarily derailed in response to plant managers' pleas for easing up on the number of corporate audits.

If, after balancing goals and objectives, resource requirements, and resource availability, you determine that all operating facilities should be audited on a regular repetitive cycle, you next need to determine the length of cycle most appropriate. (This subject is discussed in the next section.) Should you determine that a "sample" of facilities is appropriate, you need to develop criteria or a sampling scheme for selecting the facilities to be audited. (This subject is discussed in the following section.)

HOW FREQUENTLY?

In answering this question, consider whether different categories of facilities are to be included in the audit program. If so, perhaps different frequencies for different types of facilities may be appropriate.

One note of caution about varying the frequency of audits by categories or classes of facilities. Environmental audit designers and company management alike must recognize and acknowledge that the complexities of environmental risk are not that well-understood and that a single upset of an operation can lead to an adverse material impact. Thus, it is important to keep in mind when categorizing facilities for audit that significant problems could arise at a facility perceived to be lower risk and thus not selected for audit.

Appreciating this concern, many audit programs sort the audit frequency among types or categories of facilities to be included in the audit program. A simple scheme would sort the facilities into two or three categories, for example, small, medium, and large operations. Small facilities might be audited every three years, medium every two years, and large facilities annually. Other criteria that could be used for a categorization scheme are depicted below.

Examples of Criteria for Developing an Audit Frequency Scheme

Objective Criteria
    Size (and complexity) of operation
    Size of environmental staff (relative to total staff or size of operation)

Volume of emissions or effluent
Safety record
Known problems

Somewhat Subjective Criteria
Inherent environmental, health, and safety risks (relative to other facilities)
Manageable environmental, health, and safety risks (relative to other facilities)
Sophistication of environmental management systems
Suspected problems

Using these criteria as a guide, look at each of your facilities and group them roughly by each criterion. For example, group the facilities by size and complexity of operation as large (high), medium, and small (low).

Care should be exercised whenever subjective criteria are used for determining the frequency of audit. The very fact that criteria are subjective introduces the possibility of bias into the sampling scheme. Programs that attempt to focus on "high risk" facilities are vulnerable to both blind spots and under- or overestimated risks.

Lastly, in answering the question of How frequently?, it is useful to think about where audit resources should be focused. At one level of analysis, it can be argued quite convincingly that environmental audits should focus on the facilities where the problems are perceived to be greatest. However, if the objective of the audit is indeed to confirm that systems are in place and functioning and compliance is being achieved, it can be argued equally convincingly that the audit should focus at least as much on facilities that are perceived to be in good shape. Given the complexities of environmental risk and the difficulties of dealing with subjective biases, it may be useful to shift the focus from where problems are perceived to be to where the inherent potential for problems is perceived to be greatest.

SITE SELECTION

If your response to auditing all facilities on some sort of repeat cycle is no (for example, if you decide initially to audit just major production facilities and not smaller operating facilities or distribution activities), you do not need to develop an audit cycle in the typical sense. However, you will probably want to develop a process for selecting facilities to be audited.

This section briefly highlights some of the key considerations. (*Note*: Special considerations for selecting the initial one or two facilities for audit in a newly established environmental audit program are addressed in Chapter 17.)

Two types of samples can be developed: (1) a directed sample and (2) a random sample. The directed sample seeks to obtain a review of facilities selected on the basis of a particular criterion rather than attempting to obtain one representative of the entire population. An example of a directed sample would be to audit all facilities in a geographic area where regulatory inspections were thought to be lean or lacking. Another example of a directed sample would be to audit all facilities that have installed a particular piece of pollution control equipment.

A random sample attempts to eliminate any sampling bias while representing, as closely as possible, the population from which it was drawn. Certainly in the environmental, health, and safety areas, there are significant limitations in any method used to select a sample truly representative of the population from which it is drawn.

CHAPTER 19

DEVELOPING AN AUDIT REPORTING FRAMEWORK AND RECORDS RETENTION POLICY

In the preparation for and conduct of the environmental audit, numerous data, files, records, and working papers are developed and accumulated. As your audit program matures and the number of audits you conduct increases, it will become necessary to establish some type of formal policy or procedure for the maintenance and retention of audit reports and records. Chapter 19 serves as a guide for developing a basic reporting process and establishing a records retention policy appropriate for your program.

DEVELOPING A REPORTING FRAMEWORK

Chapters 7 and 14 detail issues to consider when developing an audit report and present some reporting options. In developing an effective reporting process for your program, begin by specifying the options you have chosen as they relate to your overall program goals and objectives.

Worksheet 19-1 provides a useful framework for describing or review-

WORKSHEET 19-1. AUDIT REPORTING WORKSHEET

1. List the goals and objectives of your audit program.

   Primary goal(s) _____

   _____

   _____

   Secondary goal(s) _____

   _____

   _____

2. List the objective(s) of the audit report.

   _____

   _____

   _____

3. Identify the varying needs of

   Corporate management _____

   _____

   Environmental management _____

   _____

   Facility management _____

   _____

   Corporate counsel _____

   _____

   Audit team _____

   _____

4. What is to be reported to each of the following (e.g., exceptions only, good practices, and recommendations) and when?

   Audit team _____

   _____

   Audited facility _____

   _____

WORKSHEET 19-1. *(Continued)*

Environmental staff _____

Legal staff _____

Corporate management _____

5. List the reponsibilities of those receiving the audit report.

   Primary addressee(s) _____

   Other report recipients _____

6. Develop and communicate the time requirements and responsibilities for the overall reporting process.

   Draft report _____

   Report review _____

   Final report _____

   Action plans _____

   Action plan follow-up _____

ing your reporting process. In item 1 of the worksheet, list the primary and secondary goals of your audit program. Then in item 2, list the objective(s) of the audit report. Identify the needs of those persons listed in item 3 as they relate to the audit report. Determine what is to be reported to those persons listed in item 4, and when it is to be reported. In item 5, list the responsibilities of the audit report recipients. Finally, indicate the timing and responsibilities for the overall reporting process.

PURPOSE OF A RECORDS RETENTION POLICY

Many companies have established formal policies to ensure that all records pertaining to environmental audits are retained for a period consistent with their usefulness to the audit program, applicable federal and state regulations, and corporate policy. An additional purpose, particularly after several years of audits, is to keep the volume of audit records and documents to a manageable level. By periodically updating the file to add new information and deleting outdated material, you can significantly reduce storage problems and facilitate easy retrieval.

ISSUES TO CONSIDER IN DEVELOPING A RETENTION POLICY

There are several issues to consider in developing a records retention policy. Foremost among these is the concern of having records in the file that may constitute a "smoking gun." As discussed in Chapter 14, the smoking gun risk can be minimized if the audit program is structured so that corporate policy on the treatment of noncompliance items is clear, the audit report automatically initiates corrective action, and top management is clearly apprised of the situation. To protect against this concern, advice from, and approval of, legal counsel should be sought in developing a records retention policy. The legal staff is aware of established routines for how to handle sensitive documents, and how long to retain them.

The protection of confidential documents should also be considered when developing a records retention policy. Typical methods employed for ensuring confidentiality include labeling such documents *confidential*, controlling the dissemination of such documents to only those who "need to know," limiting access to confidential documents, and keeping all confidential documents segregated from nonconfidential documents.

A third issue relates to the needs of the audit program manager. In developing a records retention policy the audit program manager should carefully examine the different types of records that are generated (e.g., working papers and audit preparation material) and the need to refer back to those records at a later date. Records need to be maintained long enough so that, if challenged, the program manager can demonstrate that a thorough audit was conducted. Similarly, the retention of certain records (e.g., plant layout diagrams) can avoid duplication of effort when facilities are audited on a routine basis.

Fourth, the retention policy for environmental audit records should be consistent with the retention policy for other records within the corporation. Usually, formal retention policies exist for accounting, internal audit, legal, and other types of records within the corporation. Similarly, audit records retention should also be consistent with retention policies of specific regulatory requirements. While there are no specific regulatory requirements pertaining to audit records, many companies have decided to keep certain audit records (e.g., final audit reports) a length of time consistent with regulatory requirements applicable to the subject area audited.

COMMON PRACTICES IN RETAINING RECORDS

Companies with established environmental auditing programs take varying approaches to retaining audit records. Some companies have no formal policy while others have formal retention policies that specify which documents are to be kept, how long they are to be kept, and where they are to be stored. Some retain all audit records for an indefinite time while others place rigid constraints on the length of time records are kept. Some companies that audit all facilities on a set schedule typically retain audit records and reports only until the next audit of that facility.

While common practice does vary, many companies with audit programs have recognized that different types of audit documents should be handled differently. Chapter 12 indicated that one way companies do this is to segregate the working papers between a "permanent file" (which includes items of continuing interest) and a "current file" (containing items useful to a specific audit). Table 19-1 lists a number of types of audit records and shows how three companies manage their record retention process.

STEPS IN DEVELOPING A RECORDS RETENTION POLICY

To develop a records retention policy for your company, use Worksheet 19-2 as a guide. First, begin by identifying and listing the different categories of audit documents such as those indicated in Table 19-1. Try to be as specific as possible. For example, list pre-audit correspondence, audit schedule, completed audit guides or protocols, and auditors notes or working papers separately rather than lumping them all into the single category of audit working papers. In addition to the audit records listed in

TABLE 19-1. EXAMPLES OF RECORDS RETENTION SCHEDULES

Document	Retention		
	Company A	Company B	Company C
Pre-audit information	Not specified	Retain and update for each audit	Retained indefinitely
Working papers (notes, exhibits, etc.)	When audit is repeated	Until final report issuance or after 90 days	Retained until action plan is issued
Draft reports	Until final report is issued	Until final report is issued	Until final report is issued
Comments on draft reports	Until action plan is received	Not specified	Not specified
Final audit report	Ten years (or longer where subject to regulations)	Until next audit	Three years
General correspondence relating to the audit	Five years	Not specified	Three years
Action plans	Ten years (or longer where subject to regulations)	Not specified	Three years
Regulations	When superceded	Not specified	When superceded

WORKSHEET 19-2.   RECORDS RETENTION WORKSHEET

1. List the specific documents and records generated.

(a.) _____

(b.) _____

(c.) _____

(d.) _____

(e.) _____

(f.) _____

(g.) _____

(h.) _____

2. Identify the specific purposes for which each of the documents might be used in the future.

	Document	Purpose
(a.)	_____	_____
(b.)	_____	_____
(c.)	_____	_____
(d.)	_____	_____
(e.)	_____	_____
(f.)	_____	_____
(g.)	_____	_____
(h.)	_____	_____

3. Make an initial determination of retention times and list the documents in order of retention times (from shortest to longest).

Document	Retention Time
_____	_____
_____	_____
_____	_____
_____	_____

WORKSHEET 19-2. *(Continued)*

Document	Purpose
_____	_____
_____	_____
_____	_____
_____	_____

4. List the key individuals necessary for developing, approving, communicating, and implementing a records retention policy.

Individual	Function
_____	_____
_____	_____
_____	_____
_____	_____
_____	_____
_____	_____
_____	_____
_____	_____

5. Identify other elements of your records retention policy relating to

   (a.) Storage of documents _____

   _____

   (b.) Key individuals responsible for

         Communicating policy _____

         Implementing policy _____

         Periodic review of policy _____

   (c.) Other _____

   _____

   _____

column 1 of Table 19-1, other audit documents may require special attention in your retention policy. For example, documents pertaining to enforcement actions or litigation should not be destroyed after an investigation has begun or if a lawsuit is pending.

Next, for each of the documents listed in step 1, identify the specific purposes for which each might be used in the future.

The third step is to make an initial cut at retention times that seem to be appropriate for each document given the needs of the audit program. The period set should be consistent with the purposes identified in step 2.

Fourth, identify the key individuals who will need to be involved in developing, approving, communicating, and implementing a records retention policy. After the "approvers" are identified, review your initial cut of retention times with each of them and modify the schedule as appropriate.

Finally, identify other elements of the records retention policy that relate to storage of records, and key responsibilities for communicating, implementing, and reviewing the policy.

After determining what you need, why you need it, and how long you need it, your records retention (and records destruction) should be routine. Dedicated files will facilitate implementing a retention policy. Periodic checks on the policy will serve as a vehicle for knowing whether the policy is followed and whether it remains appropriate to your audit program.

CHAPTER 20

MANAGING AND EVALUATING AUDIT QUALITY

As described throughout this book, a wide variety of alternatives are available to most environmental audit programs. However, not all combinations of the various alternatives are equally appropriate to a particular situation or organizational setting. Chapter 20 outlines some techniques for managing audit quality and describes some benchmarks for evaluating or critiquing your audit program.

MANAGING AUDIT QUALITY

A program for managing audit quality becomes increasingly important as internal environmental audit programs grow in size and organizational stature. Corporate management often has high expectations for the environmental audit program and such expectations usually heighten the need for an effective approach to managing audit quality.

The basic elements of a program for managing environmental audit quality are:

Performance criteria.

Supervision.

Other guidance and direction devices.

Periodic program critiques and evaluations.

Performance Criteria

The process of managing audit quality typically starts with the development—or, more often, the evolution—of informal criteria about what constitutes an acceptable audit. These performance standards tend to be specific to an individual company's environmental audit program and take into account the needs and expectations of the program's sponsors as well as the trade-offs likely to be encountered in conducting the audit. Frequently, these performance criteria are not made explicit until the audit program grows to such a size that their attainment is perceived as important.

Supervision

Supervision is the cornerstone of any program to manage environmental audit quality. Supervisors (audit team leaders, and audit program managers) should make sure that each audit is well-planned, appropriate staff and other resources are available and effectively assigned, audit program procedures are adhered to, guidelines are followed, and audit findings are solidly supported.

With respect to managing audit quality, audit supervision should:

Review the work of each audit team member to ensure that all steps of the audit protocol are completed (or documented to show why carrying out the particular audit step was inappropriate).

Review the working draft audit report in detail to ensure that report contents are fully supported by facts developed during the audit.

Approve the filing or destruction of working papers and other audit records.

Other Guidance and Direction Devices

Other audit guidance and direction devices include program policies, procedures, and guidelines. These documents are intended to provide general guidance and/or specific instructions to the audit team and other individuals. Many factors will influence the need for basic direction documents, but probably none more than program staffing and organization. An environmental audit program that is staffed with an ad hoc audit team will generally have higher needs for direction or guidance statements. Similarly, a program staffed entirely with full-time environmental auditors but organized in such a way that multiple audit teams may be in the field simultaneously may need audit procedures or guidelines to help provide continuity from audit to audit.

At one end of the spectrum, various guidance documents and instructions may be incorporated into a formal audit manual. Such a manual usually contains not only the audit protocols and questionnaires but also detailed audit instructions both for conducting the audit and for managing its quality. More common is the issuance of various procedures or guidelines as individual stand alone documents. Table 20-1 illustrates some of the more basic topics.

TABLE 20-1.   EXAMPLES OF AUDIT PROGRAM PROCEDURE OR GUIDELINE TOPICS

Working paper preparation	Example formats and other guidance for preparation of the audit work papers
Documentation requirements	Delineation of the basic documents to be collected, examined, and included in the working papers
Records retention	Criteria for retention and routine destruction of the audit records after they have outlived their usefulness
Report preparation	Identification of the roles, responsibilities, and timing of report preparation
Report review and follow-up	Description of the roles and responsibilities for reviewing audit reports, determining what action, if any, may be appropriate
Documentation of corrective action taken	Guidelines for documenting corrective action taken
Site selection and scheduling	Criteria for selection and frequency of audits; procedural checklist for scheduling the audits
External reporting and disclosure	Procedures for external reporting or disclosure of audit findings where required or deemed appropriate

Periodic Program Critiques and Evaluations

Periodic critiques or evaluations of the audit program can be especially helpful in identifying program weaknesses. A number of alternative approaches are available. The most basic (but from our experience, often one of the more difficult) is to set aside some time at or near the end of each audit for a critical self-evaluation by the audit team.

We suggest that such evaluations be structured somewhat rather than based on a single, open-ended question (e.g., How did we do?) posed all too near the appointed hour for packing up and heading home. There are many ways to structure such a self-examination by the audit team. One relatively simple approach we have found effective in eliciting the desired introspection is to ask and then answer each of three separate questions. The first question to consider is, What went particularly well during this audit? After answering this question (including noting all answers on flip charts or note pads and discussing the various responses offered by the members of the team), but only after answering the question, go onto the second question: What did not go particularly well during this audit? Again, after answering, discussing, and recording the responses, ask the third question: If we could do this audit again, what would we do differently?

A second option is to have someone critique and evaluate your program. In choosing a reviewer to critique and evaluate your audit program, be sure to consider carefully the needs for internal versus external orientation. Where your program is relatively well-developed and documented, and you desire an evaluation primarily against your own procedures and performance criteria, use of an internal group may be most appropriate. However, in situations where the evaluation should also consider current industry practices and recent developments in environmental auditing, use of individuals external to your organization may be appropriate. While most outsiders will probably have to undergo some learning period to understand your audit program, the learning curve is likely to be significantly shorter than the learning curve that employees of your firm would face because they have not been directly involved in the environmental auditing program, current environmental auditing practices, or recent developments.

An evaluation or critique by individuals at least one step removed from the program can be especially useful in identifying weak links or "blind spots" in your program's design or execution. When such evaluations are conducted by qualified individuals, they can provide comfort for the audit program manager. That is, a formal, written report resulting from a competent evaluation of the audit program can serve as a comfort statement of sorts for the program manager. Alternative approaches for such an arms-length evaluation include

Company Task Force.   Evaluation of the audit program's design and execution by company employees not directly involved in the audit program.

Performance Questionnaire.   Sent to each facility after the audit for comments, reactions, and criticisms of the audit and the conduct of the audit team.

Internal Auditors.   While often "short" on environmental expertise (keep in mind you are looking for a review of your procedures and practices, not an actual environmental audit), the internal audit department already understands auditing. Moreover, in most organizations, they already have a charter that would allow them to undertake the critique. Unless your environmental audit organization is housed in the internal audit department, this evaluation could be considered "independent" of your program.

External Peer Reviews.   An external review and evaluation can be conducted by environmental auditors from another organization, perhaps on a reciprocal basis. While not yet widely practiced within the environmental auditing community, a peer review can offer significant advantages for the reviewer as well as for the program being reviewed.

Outside Consultants.   Consultants with significant actual environmental auditing experience can provide an independent and objective evaluation of the audit program that is based on current industry practices in environmental auditing.

EVALUATING THE PROGRAM

As an initial step toward evaluating or critiquing your audit program, take a careful look at your program goals and objectives, program responsibilities and accountabilities, audit approach, staffing, and reporting.

Program Goals and Objectives

Your audit program should have a clearly defined purpose and explicit, realistic objectives. The program goals and objectives should be understood by all parties—the members of the audit team, the environmental affairs staff, those members of line mangement who receive program reports, and key managers of the audited facilities. Each should understand both the basic objectives of the program and the limitations inherent in meeting those objectives. Moreover, these groups collectively should have a shared understanding of the mission of the audit program.

To examine the effectiveness of your environmental audit program, start with a critical self-examination of your program goals and objectives. Worksheet 20-1 can provide a useful framework.

WORKSHEET 20-1. CRITIQUING YOUR AUDIT OBJECTIVES

Criteria
1. Explicit, accurate, and realistic goal definition
2. Appropriate to the situation
3. Communicated effectively to all concerned

A.   Describe your program goals and objectives. (If you have a written program policy or mission statement, you may want to review it. Otherwise, you may want to refer to your response to item B on Worksheet 15-3.)

_____

_____

_____

_____

B.   To what extent has the nature or mission of your environmental auditing effort been defined explicitly, accurately, and realistically?

_____

_____

_____

_____

C.   What steps have you taken to ensure that your goals and objectives:

1.   Have you addressed management's needs and expectations?

_____

_____

_____

2. Are they congruent with corporation's culture and management style?

_____

_____

_____

3. Are they consistent with overall environmental management philosophy?

_____

_____

_____

4. Are they consistent with (or equivalent to) audit programs of peer companies?

_____

_____

_____

D. What general steps have you taken to communicate your program goals and objectives to all concerned?

_____

_____

_____

1. What specific steps have you taken to make top management familiar with the capabilities and limitations of the audit program?

_____

_____

2. What have you done to make facility management aware of program goals so that they can ensure that audits of their facilities present a fair and accurate picture of facility environmental status?

_____

_____

3. What have you done to make sure audit team members understand program's overall mandate and their individual accountabilities?

_____

_____

4. How specifically (formal policy, memorandum, word of mouth, etc.) have the program objectives been communicated to the following?

Top management _____

Facility management _____

Audit team members _____

Others _____

Program Responsibilities and Accountabilities

In addition to explicit and realistic goals and objectives, it is also important that program responsibilities for overseeing, supervising, and carrying out the audit program be clearly defined and delineated to all concerned. Worksheet 20-2 outlines selected audit program responsibilities and provides space for you to identify the accountable individuals.

Audit Approach

The term "environmental audit" can refer to a wide variety of activities and approaches. What constitutes an acceptable audit with respect to your program? How consistent or variable is your approach from audit to audit? How consistent is your basic approach with your program goals and objectives? And, how does any variability in your program approach factor in?

Does your audit output match your goals and objectives? If your objective is to confirm compliance, does the audit diligently check key aspects of compliance? Is evidence of compliance generated and documented? Or does the audit documentation mostly focus on problems and

WORKSHEET 20-2.   CLEAR-CUT PROGRAM ACCOUNTABILITIES

Program Responsibility	Accountable Party(ies)	Method by Which Responsibility Is Assigned and Communicated
1. Program review and oversight	_____	_____
2. Updating materials	_____	_____
3. "Clearing" audit reports	_____	_____
4. Selecting audit staff	_____	_____
5. Scheduling audits	_____	_____
6. Resolving differences of opinion	_____	_____
7. Action planning	_____	_____
8. Follow-up	_____	_____
9. Quality assurance	_____	_____

shortcomings? If your objective is to confirm that management systems are in place and functioning to assure compliance, does the audit team review both the current compliance status and the key systems that are in place to manage compliance? Are those systems documented in a manner that demonstrates they are clearly understood?

Use Worksheet 20-3 to test whether your audit procedures and techniques are congruent with your program goals and whether sufficient evidence is generated and documented to support your findings.

WORKSHEET 20-3. APPROACH

A. Consider whether your audit procedures and techniques facilitate a review that is congruent with your program goals and objectives.

  1. Is a written plan prepared in advance for each audit?

      _____

  2. Are explicit instructions and objectives provided to your audit teams?

      _____

  3. If the answer to both 1 *and* 2 (above) is "No.," how do you achieve the necessary consistency given variability from facility to facility?

      _____

      _____

      _____

  4. How have you provided for periodic updating of audit materials due to

      Regulatory changes? _____

      Revisions to company standards or procedures? _____

      Learnings or developments as a result of your own experience or that of others? _____

      When were the last two times your audit materials were updated, and for what reason? _____

B. How is sufficient evidence generated and documented during the audit to to support findings in a manner consistent with program goals?

    _____

    _____

    _____

    _____

  1. Select two of your recent audit reports. Review each report and list (in the space provided below) each audit finding and observation contained in the report. (Include both positive and negative findings in your listing.)

WORKSHEET 20-3. *(Continued)*

2. For each observation noted, review the audit procedures and describe the evidence developed during the audit to support that finding. If you are unable to recreate the evidence developed, examine your audit protocols to determine what evidence, if any, is specifically called for.

3. Review work papers to determine what, if any, documentation was developed.

	Finding/Observation	Evidence Generated During Audit	Documented in Working Papers
example	*SPCC plan out of date*	*List former employee as person to call in event of spill.*	*Yes*
example	*RCRA Manifest program is outstanding*	*Informal comparison of procedures with those of four other facilities*	*Yes*
example	*Coal may not meet sulfur requirements*	*Vendor analysis not consistent with company analysis*	*No*
	_____	_____	____
	_____	_____	____
	_____	_____	____
	_____	_____	____
	_____	_____	____
	_____	_____	____
	_____	_____	____
	_____	_____	____
	_____	_____	____

Audit Team Staffing

The quality of the audit program is a direct result of the training, expertise, and proficiency of the personnel assigned to conduct the audit. In examining the effectiveness of your audit team, look carefully at whether the staffing is consistent with program goals and whether the team members have the necessary training and experience to conduct the audit. Use Worksheet 20-4 as a guide in making this assessment.

WORKSHEET 20-4. AUDIT TEAM STAFFING

A. Describe what general staffing requirements are implied by the goals and objectives of the audit program.

_____

_____

_____

_____

1. What special staffing requirements must be met because of program needs for reliability and credibility?

_____

_____

_____

2. What steps have you taken to ensure that the audit staff is large enough to provide the audit coverage desired by management?

_____

_____

_____

3. How does the staffing of the audit program take into account the program's needs for continuity and personnel commitment (e.g., full-time versus part-time auditors)?

_____

_____

_____

4. Are the organizational relationships of the auditors such that they are free of pressures or biases that may affect their objectivity? Describe.

_____

_____

_____

B. List the training and experience necessary to conduct the audit.

    1. Regulatory requirements

    2. Manufacturing operations and processes

    3. Environmental control technologies

    4. Environmental management systems

    5. Scientific disciplines

    6. State-of-the-art understanding of peer companies and facilities

    7. Relevant audit technologies

    8. Legal considerations

    9. Past experience (describe)

    10. Other

Reporting

As discussed previously, environmental auditing reports are designated to provide management with information about the environmental status, initiate corrective actions, and document the results of the audit program. A number of potentially complex issues are involved in assessing whether reporting of individual audit findings and overall program performance is appropriate. Use Worksheet 20-5 to sort through these issues.

WORKSHEET 20-5. REPORTING AUDIT FINDINGS AND RESULTS

A. How have formal reporting relationships between the audit program and various levels of management been established and specified?

_____

_____

_____

_____

1. What steps have been taken to ensure each of the following have a clear and shared understanding of what is to be reported, when, to whom, and for what purpose?

Audit team _____

Audited facility _____

Environmental staff _____

Legal counsel _____

Corporate management _____

2. How have roles, responsibilities, and channels been specified for audit findings that may require outside disclosure under existing regulatory frameworks?

_____

_____

_____

B. Are audit reports timely and concise and do they clearly and appropriately disclose audit findings to their addressees? Describe.

_____

_____

_____

_____

C. What steps have you taken to ensure the audit reports are appropriately prepared?

1. Are they factual and informative?

_____

_____

_____

2. Are any opinions and recommendations clearly labeled as such?

_____

_____

_____

3. Do the reports have a consistently applied and standardized format?

_____

_____

_____

4. Do they convey information in a medium (oral, written, combination) that is appropriate to the situation?

_____

_____

_____

5. Are they consistent with other comparable information and reports?

    _____

    _____

    _____

D. Follow-Up

    1. How do you know audit findings receive appropriate consideration and follow-up?

        _____

        _____

        _____

    2. Describe the procedures established for ensuring that

        Audit findings are carefully considered by the necessary parties.

        _____

        _____

        _____

        There is a routine review of all audit reports.

        _____

        _____

        _____

        Prompt corrective action is taken in all situations where deemed appropriate.

        _____

        _____

        _____

WORKSHEET 20-5. *(Continued)*

Appropriate steps are taken to avoid similar discrepancies elsewhere within the company.

_____

_____

_____

_____

THE REAL TEST IS IN THE RESULTS

The ultimate test of any environmental auditing program is not in the elegance of its design but in the results it produces. From what we have seen as a result of some of the more sophisticated environmental auditing programs, a soundly designed environmental auditing program, adequately staffed, and effectively implemented should result in (or at the very least help bring about)

Environmental audits that produce an accurate understanding of the current environmental status and clearly identify potential concerns within the scope of the audit.

Audit findings that are routinely reported to appropriate levels of management and that lead to timely correction of noted discrepancies and to adoption of steps to avoid repeat or similar occurrences both at the audited facility and elsewhere within the corporation.

Audit results that are borne out over time. Significant "surprises" or recurring patterns of shortcomings in environmental performance undisclosed at the time of the audit are not subsequently discovered.

Management recognition that environmental conditions are better known and understood as a result of the audit program.

Increased confidence by management that the environmental activities of the company are in good order.

An environmental audit program that has not only the support but also the confidence of top management.

PART FIVE

FUTURE OF ENVIRONMENTAL AUDITING

CHAPTER 21

EMERGING TRENDS AND FUTURE DIRECTIONS

Throughout this book, we have described environmental auditing as an emerging discipline. Since its initial beginnings in the early 1970s, environmental auditing has been practiced in different ways by different companies and, indeed, these differences in audit philosophy and approach still exist today. Given this evolutionary process, a logical question is, What does the future hold?

The range and variety of regulatory, economic, and public pressures makes it difficult, if not impossible, to predict accurately the future of environmental auditing. Nevertheless, as environmental auditing enters its second decade of practice within industry, a number of specific trends are emerging. Chapter 21 discusses these trends in four sections: (1) the driving forces behind developing and enhancing audit programs, (2) the future status of audit programs within companies, (3) emerging standard practices, and (4) the increasing focus of auditing on environmental management systems.

One important caveat should be made before discussing these issues. As with any relatively young discipline, the long-range future largely depends on the inventiveness and ingenuity of the serious practitioners and their ability to serve their "clients" effectively. Environmental auditors will undoubtedly continue to develop and refine their methods and techniques in ways not yet anticipated. Thus, Chapter 21 is not intended to "crystal ball" the future accurately. Rather, it presents current emerging trends and suggests their potential future implications.

DRIVING FORCES

We believe that the pressure for companies to develop or enhance their environmental auditing program will continue to come from a variety of places.

Top Management

In a 1981 poll, we found that more than half of those who responded indicated that top management (president, corporate counsel, etc.) and the board of directors would likely exert the greatest pressure for auditing. We believe these figures are still accurate today.

"Bottom-Up" Pressure

While top management will exert continuing—and perhaps increasing— pressure for auditing, pressure will also continue to come from the "bottom up." As environmental auditing continues to become more widely practiced in industry, environmental managers in those companies without programs will exert increasing upward pressure to sell the auditing concept. This trend will be reinforced by involvement of the environmental managers in trade associations and/or other networks of individuals who might be discussing environmental auditing.

Government Agencies

The United States Environmental Protection Agency became interested in environmental auditing during the late 1970s and by late 1981 decided to focus significant effort on exploring whether EPA and states could offer industry incentives to establish and upgrade internal environmental auditing programs. This voluntary "incentives" approach was discussed at numerous formal and informal settings and eventually determined to be too structured and programmatic, at least for the time being, for a concept as new and fluid as environmental auditing.

By late 1982, EPA shifted its attention to three interrelated initiatives: endorsement, analysis, and assistance. Endorsement refers to EPA's public support for environmental auditing, as evidenced by such activities as conducting workshops on auditing, acting as a clearinghouse to help companies learn from each other, and showcasing individual audit programs that could serve as models. As a second initiative, EPA is continuing to analyze and explore the benefits of auditing, alternative government strategies, and trends that may emerge. EPA will use the results of this analysis to assist states and companies interested in exploring joint public-private sector auditing approaches.

In 1984, EPA broadened the scope of its activities to include promotion of the environmental auditing concept to other agencies of the federal government. A few federal agencies had already initiated internal environmental auditing programs. Today, several federal agencies are developing and implementing formal programs to audit the status of their facilities. Several other agencies are actively exploring the desirability and feasibility of establishing their own environmental audit programs.

While the exact nature of EPA's future role in environmental auditing may be difficult to predict, its involvement has prompted, and we believe will continue to induce, companies and other organizations to examine carefully the merits of an auditing program. EPA's interest has helped to focus attention on environmental auditing. Additionally, the Agency's Regulatory Reform Staff has initiated several projects intended to help those interested better understand environmental auditing.

FUTURE STATUS

The continued pressure for environmental auditing from top management, middle management, and government agencies suggests that in the future, environmental auditing programs will, in the authors' opinions, be characterized by more independence, a higher organizational status, more documentation, and a higher reporting level.

Independence

As the discipline of environmental auditing continues to mature, the expectations of management will become higher, resulting in a need for demonstrating greater independence on the part of environmental auditors. Independence here refers to auditors being able to carry out their work freely and objectively, so that they are in a position to provide management with an impartial and unbiased judgment of the facility's environmental status. We expect steps will be taken within companies to increase the independence of the audit team members and the program manager by ensuring that they are outside the line of responsibility for the facility they are auditing. In situations where independence is particularly important, companies will continue to use third parties.

Organizational Status

Environmental auditing programs often begin as one of many activities an environmental staff person must manage. Very often the audit program manager has other responsibilities relating to providing staff assistance to the facilities. As more and more companies develop audit programs and as the existing programs become more sophisticated, the organizational status of the programs is expected to increase. This increase may be in "real" terms (e.g., a higher level program manager), but is more likely to be in relative terms (e.g., more responsibility, more high-level exposure, etc.).

Documentation

The increasing need for documentation stems from the need to be able to demonstrate that the environmental auditing program is, in fact, accomplishing what it set out to do. We believe that this increase will focus as much, if not more, on the quality of what is written down than the quantity. The objective is clearly not to generate more paper; rather, it is to have the documentation available to show that the audit team did a comprehensive audit, that the findings are explicit, accurate, supported by audit evidence, and understood, and that corrective action is promptly initiated and properly managed.

Reporting Level

Top management's growing interest in environmental auditing will continue to result in their desire to be apprised of the status of the program on a periodic basis. Neither top management nor the board want to know all of the details of an individual audit, but they do want enough information to be assured that the program is meeting their needs.

DEVELOPMENT OF STANDARD PRACTICES

When EPA abandoned its voluntary "incentives" approach, it did so because standard environmental auditing practices simply did not exist at the time, nor did it appear that they were likely to in the near future. Nevertheless, as more companies and public sector organizations develop programs, as the body of auditing literature increases, and as auditing concepts are discussed in broader circles, elements of standard practices will begin to emerge. Whereas during the past five years, a variety of seminars and conferences have focused primarily on those that have audit programs explaining the basics of environmental auditing to those without programs, some (perhaps even many) of the future gatherings will be comprised of individuals experienced in auditing focusing on details rather than the basics.

Several factors are (and will likely continue) contributing to the emergence of standard practices. First, as companies develop audit programs and complete the initial round(s) of audits, they often look for ways to compare notes with others in a similar position. In recent years, a number of informal networks and ad hoc groups have been formed to fill this need. Second, the continuing interest by EPA, state agencies, and others in acting as a clearinghouse of ideas suggests that it will become easier and easier for environmental audit program managers to identify those elements that are common among the more sophisticated auditing programs.

While EPA's involvement will make more information available to those who are interested in environmental auditing, the group most likely to accelerate the development of standards and procedures is audit program managers, rather than regulatory agencies or industry associations. Considerable diversity of practice will remain, and it will be the program managers from individual companies who sift through that diversity to identify practices that best meet their needs.

INCREASING FOCUS ON ENVIRONMENTAL MANAGEMENT SYSTEMS

The needs of top management and the complexities of environmental management suggest that future audit programs will shift their focus toward auditing the environmental management systems rather than the facility's compliance status at the time of audit. Top management needs to be assured that the facility has the programs and procedures in place to manage for compliance before, during, and after the audit—not just that the facility was in good shape during the audit.

The pressure for an increasing focus on systems in place at the facility also comes from environmental managers. Environmental and line managers alike are faced with three important questions.

Are the environmental management systems in place?

Can the same management system reliability and control be achieved for less effort?

In addition to knowing that appropriate management systems exist, can it be *demonstrated* that the systems are functioning properly?

Regardless of the diversity of current practice, virtually all environmental auditing programs were established to develop a systematic process for determining environmental compliance status, to use an effective methodology for collecting information, to cause corrective action to be initiated if required, and to communicate information on the company's environmental status to management. As environmental auditing moves beyond the embryonic stage and audit programs begin to mature, an increasing focus on environmental management systems should achieve the program objectives more effectively.

SUMMARY: AN EMERGING DISCIPLINE

In summary, environmental auditing is maturing to a stage where it has begun to assume many of the characteristics of a practice area or discipline.

A Body of Knowledge. There is a growing body of knowledge based upon first-hand experience with auditing, supported by an increasing volume of literature devoted to the subject.

An Underlying Rationale or Logic System. Despite the diversity of current practice, there is a common thread of examining environmental status among those companies with programs and an emerging common methodology or logic system behind how to audit.

Often Competing Theories. Companies conduct audits in very different ways to fulfill different needs. Differences in breadth and depth of audit, resource commitment, staffing and reporting will continue to dominate the field.

Basic Research.   A variety of groups have begun to do research in the discipline of auditing. In addition, universities have begun to offer courses that focus, at least in part, on environmental auditing.

Learned Societies and Journals.   While a journal devoted entirely to auditing does not exist, auditing is gaining considerable attention in a number of environmental, health, and safety journals, magazines, and other publications. Several professional associations and societies have begun to devote portions of their publications and programs to environmental auditing. Additionally, publication of a newsletter devoted to environmental auditing has recently been initiated.

Practitioners with Tools.   As companies conduct more audits, they develop more sophisticated approaches and an established framework or set of tools that is used consistently from audit to audit.

Five years ago, environmental auditing could not be characterized as having many of these elements. Today the discipline is in a rapid growth phase. We believe that five years from now, while diversity will remain, many of these elements of a discipline or practice area are likely to have taken considerable shape.

APPENDIXES

APPENDIX A

WATER POLLUTION CONTROL AUDIT PROTOCOL

This audit protocol is intended as a guide for designing and conducting facility audits. It may require additions, revisions, or other modifications in order to meet the needs of your particular audit objectives, industrial setting, or other special circumstances.

Facility Name: _____ Date(s) of Audit: _____

Team Members Participating
in Water Pollution Control Audit: _____

Protocol   Protocol
Prepared By: _____   Reviewed By: _____

	Auditor(s)/ Comments	Working Paper Reference (List Page Numbers)
1. Using data supplied by facility or provided by the audit team leader, review the following:		

	Auditor(s)/ Comments	Working Paper Reference (List Page Numbers)

a. Manufacturing process flow diagrams and descriptions

b. Facility layout including sewer diagrams and wastewater treatment system flow diagrams

c. Regulations—federal, state, and local (including present state of water quality parameters for receiving bodies)

d. Any outstanding court orders: variances, compliance orders, administrative orders, etc.

e. Applicable corporate policies, procedures, and standards

f. Facility policies and procedures

g. Spill control plans

h. NPDES and other water pollution control applications and permits

i. Quality assurance and reliability testing program for wastewater analyses

j. Status of priority pollutant assessment program

2. Review the Water Pollution Control Questionnaire completed during the advance visit. Familiarize yourself with facility responses. (If questionnaire was not administered in advance, complete Water Pollution Control Questionnaire.) Make inquiries as desired to increase your understanding of the facility's responses.

	Auditor(s)/ Comments	Working Paper Reference (List Page Numbers)

3. At some time during the audit, tour the facility following manufacturing processes and spill control plans.

    a. Determine if process flow diagram, sewer diagram, and facility layout are consistent with observed discharge locations.

    b. Determine through inspection that all discharges are conveyed to a point source location designated in permits.

    c. Based on observation, note any locations where it appears that process drainage leaves the facility site.

    d. Based on inquiry, review of sewer and process flow diagrams, and observations, determine that separate storm sewers do not convey process wastewaters or storm water run-off contaminated by contact with raw material wastes or pollutant-contaminated areas.

    e. Determine that all flow rate measurements and water sampling and monitoring devices are downstream of all discharges specified on permits.

    f. For items (c), (d), and (e) above, prepare your conclusions in narrative form.

    g. Utilizing flow diagram, facility layout maps, sewer diagrams, etc., identify and locate all oil storage and handling facilities.

	Auditor(s)/ Comments	Working Paper Reference (List Page Numbers)

h.   Record the location and capacity of all oil storage tanks and determine whether based on storage capacity an oil SPCC plan is required.

i.   Determine and record the location of all materials requiring dikes and determine if tank farm separation and diking requirements are consistent with federal, state, and local requirements and company policy.

j.   Record location of all oil and other liquid transfer equipment used in loading and unloading.

k.   Record location of all spill monitoring and control equipment.

l.   Note any location where spillage would appear to create an event or where past or present spillage results in leakage to receiving waters.

m.   Determine the location of equipment for preventing spill events, especially for barge loading and unloading facilities.

4.   Document in flow chart or narrative form, your understanding of the following:

a.   Water pollution control systems and procedures

b.   Monitoring and sampling programs and procedures

c.   Procedures for record keeping and reporting of compliance

	Auditor(s)/ Comments	Working Paper Reference (List Page Numbers)

d.   Testing and analytical procedures

e.   Information handling and documentation for (b) through (d) above

5.   Prepare a schedule of requirements under outstanding court orders (see step 1) and verify that all requirements are being met.

Water Pollution

6.   Utilizing the permits or applications for the period under review (i.e., NPDES, SPDES, POTW, pretreatment, etc.) perform the following:

a.   Permit applications

(1)   Examine for completeness of data, accuracy, and compliance with required filing dates.

(2)   Document that the permit application was signed by the person specified in regulatory rules or that person's duly authorized representative.

(3)   Determine through review of capital investment files, facility production records, or operating budgets that expansion of facilities or increases in production rates are properly reflected in applications for revisions or or renewal of permits. Also, determine if any process changes

	Auditor(s)/ Comments	Working Paper Reference (List Page Numbers)

may have resulted in changes of type of pollutants not covered in applications for permits or in existing permits.

(4) For those substances reported as present or absent in discharges at time of permit application, ascertain if the basis for that description is still valid.

(5) Ascertain that if any water quality parameters for receiving waters would necessitate revisions in permit applications.

b. Permits

(1) Develop a schedule of all permit requirements currently in effect.

(2) Based on flow diagrams and observations, determine that operating procedures or installed systems are capable of providing information substantiating compliance with requirements.

(3) Ascertain status of compliance with all stipulations of permits.

(4) For the review period, record date of all known excursions, type of excursion, abatement or corrective actions, and communications to regulatory agencies, etc. Determine if compliance with all reporting procedures has been carried out; if not, list exceptions and mitigating circumstances.

	Auditor(s)/ Comments	Working Paper Reference (List Page Numbers)

(5) Obtain and review intracompany correspondence relative to permit limitation issues or regulatory actions, noting all unresolved issues.

7. Develop a flow chart or other description showing process and responsibilities for water pollution control activities (e.g., responsibility for sampling, analysis, record maintenance, and regulatory reporting).

8. Based on information previously developed and your understanding of the system, confirm the operation of the data collection and reporting system.

a. Data collection

(1) With facility personnel, observe the procedure for sample collection, analysis, and data recording. Document the maintenance and calibration programs for composite sampling, effluent flow measuring, in-place monitoring and recording devices, and control equipment.

(2) Note any provisions for cross-checking or verification by independent analysis.

(3) Assess appropriateness of maintenance and calibration programs:

(4) Ascertain that approved analytical techniques are utilized and that maintenance of laboratory instruments is routinely performed.

	Auditor(s)/ Comments	Working Paper Reference (List Page Numbers)

b. Test of reporting procedures

   (1) For each active permit, select a day of noncompliance with permit limitations (if appropriate) and select a day of compliance with permit limitations.

   (2) Obtain documents for days selected in (1) and carry out the following procedures:

   Review recorder charts, laboratory results of the permit required sampling, etc.

   For the noncompliance day, review file material to ascertain basis for noncompliance incident report.

   Review reports forwarded to regulatory agency confirming their completeness, accuracy in submittal within specified time period, etc;

   Review internal records for the proposed corrective action.

   Determine that corrective action was effected promptly (i.e., revision of operating procedures, repair of equipment, installation of new equipment, etc.).

9. Review plans and programs for compliance with published regulatory requirements and confirm that dates of compliance are consistent with requirements.

10. Prepare detailed list of review findings.

	Auditor(s)/ Comments	Working Paper Reference (List Page Numbers)

Spill Prevention and Control

11. Based on review of SPCC plan, determine its compliance with mandated submittal time, certification by licensed professional engineer, implementation, review, requests for extention of time and completeness for consideration of all oils and oil-containing substances as well as other liquids.

12. Develop a flow chart showing responsibilities for action of record keeping for spill prevention and control (e.g., responsibility for initiating corrective actions, regulatory reporting, and record maintenance).

13. Based on observations made during facility tour, determine if all preventative systems are capable of preventing spills from entering surface waters. If appropriate preventative systems are not installed, review and document the rationale supporting the absence of such systems.

14. Document in writing your understanding of the drainage system on the facility site and, for selected diked storage areas or other passive retention systems, determine through inspection that the systems are operating as designed. In particular, determine the effectiveness of valves and operating procedures to retain spills (e.g., confirm that provisions are adequate to prevent accumulated rain water), from decreasing the design volume of diked areas.

	Auditor(s)/ Comments	Working Paper Reference (List Page Numbers)

15. Determine that drainage directly into water courses or into catchment basins are inspected and records kept of inspections. Also, ascertain that catchment basins are not subject to periodic flooding.

16. From engineering specifications and purchasing specifications, determine that materials and construction of tanks are compatible with materials stored and conditions of storage.

17. Ascertain that the area within dikes is sufficiently impervious to contain spills.

18. For underground metallic storage and piping installations, ascertain that appropriate corrosion protection has been installed (e.g., cathodic protection and resistant coatings).

19. Review facility maintenance and test records on storage tanks, tank supports and foundations, and valves and piping. (*Note:* This information may be obtained from work orders or inspection reports, maintenance schedules, etc.) Particularly review records to reflect that significant trends had been detected and appropriate corrective action taken.

20. For tanks heated internally by steam, document the manner in which condensate is handled.

21. For selected tanks with level-sensing devices, physically examine such devices and

	Auditor(s)/ Comments	Working Paper Reference (List Page Numbers)

review maintenance records for frequency of testing.

22. For piping not in service, determine if the terminal connections are properly capped.

23. For loading and unloading facilities, determine adequacy of signs and procedures for warning vehicular traffic, ensuring that vehicles cannot depart before complete disconnect; ascertain that catchment basin serving the area is adequate to hold at least the maximum capacity of any single compartment of a tank truck or car.

24. Inspect the procedure by which oil or other liquid pumps are controlled and ascertain if operations could be performed by unauthorized personnel.

25. Record frequency of spill prevention briefings, attendees, and records of meetings.

26. If the facility has discharged more than 1,000 gallons of oil or other liquid in a single spill event or discharged in "harmful" quantities two spills within any 12 months, confirm that a report was submitted to the appropriate regulatory agencies within the required 60 days. Confirm that corrective actions required in the report were implemented. Examine all correspondence between company and regulatory agencies, noting particular reference to compliance status.

27. Prepare detailed list of audit findings.

APPENDIX B

WATER POLLUTION CONTROL QUESTIONNAIRE

This audit questionnaire is intended as a guide for designing and conducting facility audits. It is intended to be used in conjunction with an audit protocol. This questionnaire may require additions, revisions, or other modifications in order to meet the needs of your particular audit objectives, industrial setting, or other special circumstances.

A. Water Pollution

	Yes	No	N/A
1. Which of the following activities are conducted?			
a. Treated wastewater is discharged to surface waters.	___	___	___
Permitted?	___	___	___
b. Untreated process wastewater is discharged to surface waters.	___	___	___
Permitted?	___	___	___

	Yes	No	N/A

c. Surface run-off.   ⎯⎯ ⎯⎯ ⎯⎯

    Permitted?   ⎯⎯ ⎯⎯ ⎯⎯

d. Discharges to POTW.   ⎯⎯ ⎯⎯ ⎯⎯

    Permitted?   ⎯⎯ ⎯⎯ ⎯⎯

e. UIC (deep well) injection (on-site or off-site).   ⎯⎯ ⎯⎯ ⎯⎯

    Permitted?   ⎯⎯ ⎯⎯ ⎯⎯

f. Ocean dumping.   ⎯⎯ ⎯⎯ ⎯⎯

    Permitted?   ⎯⎯ ⎯⎯ ⎯⎯

g. On-site disposal (e.g., percolating beds, irrigation).   ⎯⎯ ⎯⎯ ⎯⎯

    Permitted?   ⎯⎯ ⎯⎯ ⎯⎯

2. Have all required applications been filed with appropriate authorities at the state, federal, and local levels, and are all permits currently valid?   ⎯⎯ ⎯⎯ ⎯⎯

3. Were all applications for permits signed by authorized company officials?   ⎯⎯ ⎯⎯ ⎯⎯

4. Does the facility comply with all terms, schedules, and other requirements of the required permits?   ⎯⎯ ⎯⎯ ⎯⎯

5. Is (or was) the facility on a compliance schedule prior to specification of regulatory limits?   ⎯⎯ ⎯⎯ ⎯⎯

    a. Were the specified dates met?   ⎯⎯ ⎯⎯ ⎯⎯

    b. Were required reports submitted in accordance with schedules?   ⎯⎯ ⎯⎯ ⎯⎯

6. Have there been changes that would result in changing the terms of permits, and have these been reported to the appropriate agency? For example,

	Yes	No	N/A

a. Facility expansion, modification, or shut down ___ ___ ___

b. Production increase, modification, or decrease ___ ___ ___

c. Process modification ___ ___ ___

d. Quantity and type of pollutants ___ ___ ___

e. Establishment of water quality parameter for receiving waters ___ ___ ___

7. Have the facility's existing permits been revoked, suspended, or modified by the issuing agencies at any time since issuance? ___ ___ ___

8. Have appropriate renewal applications been filed in a manner consistent with regulatory requirements? ___ ___ ___

   a. Does the facility maintain files on all information resulting from permit monitoring activities? ___ ___ ___

   b. Is the information of (a) filed at other locations in the corporation (e.g., corporate headquarters)? ___ ___ ___

   c. Does the facility retain records for a least three years on all monitoring activities and results, including original dated strip charts, calibration, and maintenance records? ___ ___ ___

9. Does the facility have a separate storm sewer system? ___ ___ ___

10. Are any of the following wastewaters discharged into a separate storm sewer?

    a. Process wastewater ___ ___ ___

    b. Storm water from raw materials storage, process areas, pollution contaminated soils, etc. ___ ___ ___

	Yes	No	N/A

c. Sanitary wastewaters   ___ ___ ___

11. Are there any unpermitted discharges from point
     sources into surface waters?   ___ ___ ___

   a. Does all process wastewater and draining
         from manufacturing areas discharge from the
         point sources that have been permitted?   ___ ___ ___

   b. Are there any leachates from present or pre-
         vious landfills, process operations, or storage
         areas entering surface waters or likely to
         leave the premises in underground water
         flows?   ___ ___ ___

   c. Have samples been obtained of these leach-
         ates and analyzed for pollutants of concern?   ___ ___ ___

12. Does incoming water contain pollutants specified
     in permits?   ___ ___ ___

   a. If so, is the incoming water periodically
         analyzed for these pollutants?   ___ ___ ___

   b. Are permits based on these incremental
         additions from the plant?   ___ ___ ___

13. Does the discharge from the facility contain any
     detectable amounts of the following regulated
     toxics?   ___ ___ ___

   a. Aldrin or dieldrin   ___ ___ ___

   b. DDT, DDD, or DDE   ___ ___ ___

   c. Endrin   ___ ___ ___

   d. Toxaphene   ___ ___ ___

   e. Benzidine   ___ ___ ___

   f. Polychlorinated biphenyls   ___ ___ ___

	Yes	No	N/A

14. Has the facility complied with the conditions, requirements, and terms for toxic pollutants as required in 40 CFR, Part 129? ____ ____ ____

15. Have any excursions beyond permitted discharge levels occurred within the review period? ____ ____ ____

    a. If so, have more than four excursions occurred? ____ ____ ____

    b. Have the appropriate agencies been notified (e.g., five-day letters)? ____ ____ ____

16. Does the company and/or its contract laboratory utilize analytical procedures for the analysis of pollutants in compliance with permit requirements or as outlined in 40 CFR, Part 136? ____ ____ ____

    a. If no, have alternate test procedures been approved by appropriate regulatory authorities? ____ ____ ____

17. Is the inventory of discharge values included in the application for NPDES or SPDES permit still accurate? ____ ____ ____

18. Have samples been obtained for analyses of plant effluents for the 129 priority pollutants? ____ ____ ____

    a. Were these analyses performed by:

        Company laboratory? ____ ____ ____

        Outside laboratory? ____ ____ ____

        EPA laboratory? ____ ____ ____

19. Are all meetings with, visits by, or understandings with regulatory agencies documented (e.g., approvals for variances, future actions)? ____ ____ ____

	Yes	No	N/A

B. Spill Prevention and Control

20. Does the facility have an approved SPCC plan? ___ ___ ___

21. Was the plan effective by July 10, 1974? ___ ___ ___

22. Was the facility's SPCC plan implemented not later than July 10, 1975? ___ ___ ___

23. Has the current plan been reviewed and certified by a licensed professional engineer? ___ ___ ___

24. Does the plan identify critical surface water and use areas to facilitate reporting oil discharges and/or chemical discharges? ___ ___ ___

25. Is there a readily available list of names, telephone numbers, and addresses of company persons and alternates to receive notification of any discharges? ___ ___ ___

26. Is there an established (published) procedure for reporting these discharges to appropriate regulatory agencies? ___ ___ ___

27. Have prearrangements been made for requesting assistance in the event of a major spill? ___ ___ ___

28. Have provisions been made to assure that the full resource capabilities are known and can be committed in the event of a discharge? ___ ___ ___

29. Have identification and inventory of local and regional equipment, materials, and supplies been prepared for coping with a spill? ___ ___ ___

30. For the maximum credible discharge, has an estimate been made of equipment, materials, and

	Yes	No	N/A

supplies required for containment and cleanup of a spill? — — —

31. Have more than 1,000 gallons of oil been discharged from the facility site into adjoining surface waters in a single-spill event subsequent to January 10, 1974? — — —

32. Have two or more spill events occurred within any 12-month period subsequent to January 10, 1975 that were considered to be of harmful quantities? — — —

  a. Was the EPA regional administrator provided with an informational report within 60 days of the spill? — — —

  b. Was the above report also sent to the state agency in charge of water pollution control? — — —

  c. Did any regulatory agency conduct an on-site review after the above occurrence? — — —

  d. As a result of regulatory or company reviews, were any recommendations made of changes or procedures, types of equipment, or other requirements necessary to prevent and contain oil discharges? — — —

  e. Has any regulatory agency required the SPCC plan to be revised due to reportable discharges? — — —

33. Is the SPCC plan approved by management at the level with authority to commit necessary resources? — — —

34. Does the SPCC plan include a prediction of direction, rate of flow, and total quantity of oil that can be discharged? — — —

	Yes	No	N/A

35. Does the facility employ any of the following systems to prevent discharged spills from reaching surface waters?

  a. Dikes, berms, or retaining walls sufficiently impervious to contain spilled oil ___ ___ ___

  b. Curbing ___ ___ ___

  c. Gutters or other draining systems ___ ___ ___

  d. Weirs, booms, or other barriers ___ ___ ___

  e. Spill diversion ponds ___ ___ ___

  f. Retention ponds ___ ___ ___

  g. Sorbent materials ___ ___ ___

36. Does the facility have an oil spill contingency plan with written commitment of manpower, equipment, and materials to contain and remove the oil discharge? ___ ___ ___

37. Is the drainage from diked storage areas controlled or monitored to prevent a spill or excessive leakage from entering the drainage system or effluent treatment system? ___ ___ ___

38. Does the facility drainage flow into any of the following? ___ ___ ___

  a. Surface waters ___ ___ ___

  b. Wastewater treatment facility ___ ___ ___

  c. Ponds, lagoons, or catchment basins designed to retain oil ___ ___ ___

39. Are catchment basins located in areas where periodic flooding may occur? ___ ___ ___

40. Does the facility have a diversion system that could return oil to the facility from the final discharge of all in-plant ditches? ___ ___ ___

	Yes	No	N/A

41. Are there any treatment units used for the treatment of drainage waters? — — —

42. Are any oil storage tanks used that are not compatible with the material and conditions of storage (e.g., pressure and temperature)? — — —

43. Does the facility have any buried or partially buried metallic storage tanks? — — —

    a. Are these tanks corrosion protected? — — —

    b. Are these tanks subjected to periodic pressure tests? — — —

44. Do maintenance schedules require periodic integrity testing of above-ground tanks? — — —

45. Do the tanks utilize internal heating coils? — — —

    a. Do the discharges of these coils flow into surface waters? — — —

    b. Is this discharge monitored? — — —

46. In an effort to detect spills or leaks, are the tanks equipped with any of the following? — — —

    a. High-liquid level alarms — — —

    b. High-liquid level cutoff devices — — —

    c. Direct audible or code signal communications — — —

    d. Fast-response system for determining the liquid level of each tank — — —

47. Are regulatory scheduled tests made on liquid-level sensing devices? — — —

48. Does the facility have any buried piping installations? — — —

		Yes	No	N/A

49. Does the facility have any piping not in service? ___ ___ ___

50. Are all above-ground valves and pipelines regularly tested? ___ ___ ___

51. Can vehicular traffic enter facility areas where it could endanger above-ground piping? ___ ___ ___

52. Does the rack area drainage flow into a treatment facility or a catchment basin? ___ ___ ___

53. To prevent vehicular departure before complete disconnect of transfer lines, are any of the following devices used? ___ ___ ___

    a. Interlocked warning light ___ ___ ___

    b. Physical barrier system ___ ___ ___

    c. Warning signs ___ ___ ___

54. Does the facility have a procedure to determine that the lower-most drain and all outlets are closed prior to filling and departure of any tank truck? ___ ___ ___

55. Is the facility fenced? ___ ___ ___

56. Are the master flow and drain valves locked in the closed position while not operating or on a stand-by status? ___ ___ ___

57. Are the starter controls on all pumps locked in the off position while pumps are not operating? ___ ___ ___

58. Does the facility have a personnel training program for the operating and maintenance of equipment? ___ ___ ___

	Yes	No	N/A
59. Does the facility have a person designated who is accountable for oil spill prevention and who reports to line management?	___	___	___
60. Are there written procedures for inspections and examinations required by 40 CFR 112?	___	___	___
a. Are these procedures and records of inspection incorporated into the SPCC plan?	___	___	___
b. Are these procedures and records maintained for at least three years?	___	___	___

Audit team member(s) completing questionnaire _____

_____

Location personnel providing responses _____

_____

APPENDIX C

AIR POLLUTION CONTROL AUDIT PROTOCOL

This audit protocol is intended as a guide for designing and conducting facility audits. It may require additions, revisions, or other modifications in order to meet the needs of your particular audit objectives, industrial setting, or other special circumstances.

Facility Name: _____ Date(s) of Audit: _____

Team Members Participating
in Air Pollution Control Audit: _____

Protocol   Protocol
Prepared By: _____ Reviewed By: _____

	Auditor(s)/ Comments	Working Paper Reference (List Page Numbers)
1. Using data supplied by the facility or provided by the audit team leader, review the following:		

	Auditor(s)/ Comments	Working Paper Reference (List Page Numbers)

a. Manufacturing process flow diagrams and descriptions including control devices

b. Facility layout, including location of all stationary sources

c. Emission inventory

d. Regulations—federal, state, and local (including attainment/nonattainment status for each pollutant)

e. Any outstanding court orders—variances, compliance orders, administrative orders, etc.

f. Applicable corporate policies, procedures, and standards

g. Facility policies and procedures

h. Operating manuals for air pollution control systems

i. Air pollution alert and emergency plans

j. Operating and construction or modification applications and permits

2. Review the Air Pollution Control Questionnaire completed during the advance visit. Familiarize yourself with facility responses. (If questionnaire was not administered in advance, complete Air Pollution Control Questionnaire.) Make inquiries as desired to increase your understanding of the facility's responses.

3. At some time during the audit, tour the facility following manufacturing processes and facility layout of stationary sources.

	Auditor(s)/ Comments	Working Paper Reference (List Page Numbers)

a. Using the process flow diagrams and facility layout, locate and note all points of continuous and periodic emissions.

b. Where possible, list pollutants in emissions.

c. Record location of all emission control and monitoring facilities.

d. Note any likely sources of emissions not included above (e.g., fugitive dust and process losses).

e. Note location and type of activity of the facility's neighbors (e.g., industrial, residential, and institutional).

4. Document in flow chart or narrative form, your understanding of the following:

a. Air pollution control systems and procedures

b. Continuous monitoring

c. Periodic monitoring and sampling programs and procedures

d. Procedures for record keeping and reporting of compliance

e. Testing and analytical procedures

f. Information handling and documentation procedures for (b) through (e) above

5. Prepare a list of all active, pending, and inactive regulatory permits, operating certificates and registrations for airborne emissions (during the period under review). In-

	Auditor(s)/ Comments	Working Paper Reference (List Page Numbers)

clude notation of agency, facility, type of permit, effective date, expiration date.

   a. Confirm that all identified emission sources are covered by permits.

   b. Review for completeness the applications for registration, certification, operation, and construction.

   c. Determine that applications are accurate and signed by appropriate company personnel

   d. Check compliance with Prevention of Significant Deterioriation and nonattainment rules.

6. Develop a schedule of capital expenditures made during the period under review, and confirm that appropriate permits were obtained for all situations in which they were required.

7. Using the emission sources developed in step 3, and the permit data developed in step 5,

   a. Determine that all applicable standards, limitations, and compliance schedules are being met for regulated emissions from all sources.

   b. Determine through inquiry, observation, and review of available documents that the facility is in compliance with applicable opacity standards.

   c. Determine that emission and fuel monitoring and test methods and pro-

	Auditor(s)/ Comments	Working Paper Reference (List Page Numbers)

cedures followed are in accordance with applicable regulations.

    d. Review supporting documentation noting that all instrumentation, analytical techniques, calculations, records, and sampling locations and methods are in accordance with state and local regulations.

8. For those facilities with requirements under CFR 40, Part 60.7 (new source performance standards), complete the following:

    a. For the period under review, examine regulatory required monitoring reports submitted in compliance with CFR 40, Part 60.7 (new source performance standards).

    b. Review emission and ambient air monitoring records for completeness in accordance with CFR 40, Part 60.7 (d) (new source performance standards). The review should include at least

        (1) Determination that performance tests were in accordance with CFR 40, Part 60.8.

        (2) Review of maintenance department records for calibration, reliability checks, and adjustments of monitoring devices.

        (3) Investigation and documentation of any unusual items noted during the file review.

    c. Review facility's monitoring systems noting compliance with CFR 40, Part

	Auditor(s)/ Comments	Working Paper Reference (List Page Numbers)

60.13. The review should include at least the following:

(1) Determine that monitoring systems and devices were installed and operational *prior* to conducting performance tests.

(2) Confirm that performance specifications noted in Appendix B of CFR 40, Part 60 are met.

(3) Determine that the manufacturer's written requirements or recommendations for checking the operation or calibration of the devices were carried out.

(4) Document the adequacy of any alternative monitoring requirements used in the absence of a continuous monitoring system.

9. For the latest National Emission Standard for Hazardous Air Pollutants (NESHAP), determine that permit or standard limitation is being met.

a. Establish that approved testing and reporting procedures are being complied with, especially for

(1) Asbestos

(2) Beryllium

(3) Mercury

(4) Vinyl chloride

(5) Lead

(6) Benzene

(7) Acrylonitrile

	Auditor(s)/ Comments	Working Paper Reference (List Page Numbers)

10. Review hazardous air pollutant source documents applicable to construction, modification, or operation of stationary sources. Check the following:

   a. Timeliness with which the information was filed

   b. Completeness of the information forwarded to the regulators

   c. Approvals and/or subsequent correspondence from regulators (i.e., waiver of compliance)

11. Review the facility's air pollution alert and emergency plan and note any significant departures from regulations and determine that they are up to date.

12. Review a listing of public nuisance complaints (e.g., odors, dust) for period under review.

13. Prepare a schedule of requirements under any outstanding court orders (see step 1) and verify that all requirements are being met.

14. Confirm that the sulfur content of fuel oil used for fuel-burning equipment is less than that prescribed in applicable regulations.

15. Determine that the fuel dispensed for motor vehicles is in compliance with applicable federal regulations.

	Auditor(s)/ Comments	Working Paper Reference (List Page Numbers)
16. Review questionable items with responsible facility personnel. Note explanations and disposition of such items.		
17. Prepare a detailed list of audit findings.		

APPENDIX D

AIR POLLUTION CONTROL QUESTIONNAIRE

This audit questionnaire is intended as a guide for designing and conducting facility audits. It is intended to be used in conjunction with an audit protocol. This questionnaire may require additions, revisions, or other modifications in order to meet the needs of your particular audit objectives, industrial setting, or other special circumstances.

		Yes	No	N/A
1.	Are permits in effect for all required emission sources?	___	___	___
2.	Has there been any construction and/or modification of the stationary sources within the time limits of the permits?	___	___	___
3.	Were regulatory agencies notified of any construction and/or modification changes?	___	___	___

	Yes	No	N/A

4. Does the facility maintain records indicating the occurrence and duration of any malfunctions during start-up, operation, or shutdown? —— —— ——

5. Is a written report of excess emissions submitted to the appropriate agency at least quarterly? —— —— ——

6. Was the facility required to install any continuous monitoring systems? —— —— ——

7. Has the facility been required to conduct any performance tests of air pollution control systems? —— —— ——

    a. If yes, was the test conducted under conditions specified by the regulatory agency? —— —— ——

    b. Was the regulatory agency notified 30 days prior to the test? —— —— ——

8. Is the facility located in an area requiring prevention of significant degradation? —— —— ——

9. Is the facility required to comply with a specified opacity standard? —— —— ——

10. Are there any sources of visible emissions that might be concealed by facility construction? —— —— ——

11. Are permanent files maintained at the facility for the following?

    a. Emission monitoring —— —— ——

    b. Ambient air monitoring —— —— ——

    c. Calibration checks of monitoring devices —— —— ——

    d. Measurements for determining performance —— —— ——

    e. Records of maintenance of monitoring devices —— —— ——

    f. At least two years following the dates of execution of calibration, monitoring, etc. —— —— ——

	Yes	No	N/A
12. Are any of the following present in significant amounts in facility emissions?			
a. Asbestos	___	___	___
b. Beryllium	___	___	___
c. Mercury	___	___	___
d. Vinyl chloride	___	___	___
e. Benzene	___	___	___
f. Trichloroethylene	___	___	___
13. Have procedures been established consistent with regulations for the proper handling of asbestos during demolition operations?	___	___	___
14. In the case of construction and/or modification, has the appropriate regulatory group approved the facility's application for air emission permits?	___	___	___
15. Has there been any change in air emission sources after the effective date of standards outlined in CFR 40, Part 61 that has required notification be submitted to regulators?	___	___	___
a. If so, were submittals in compliance with established dates?	___	___	___
16. Does the facility operate under any waivers of compliance from a regulatory agency?	___	___	___
17. Are all emission testing and compliance monitoring in accordance with CFR 40, Part 61 and/or permit specifications?	___	___	___
18. Has the facility had any emissions in excess of permit limits that required reporting to the applicable regulatory agency?	___	___	___

	Yes	No	N/A

19. Is the facility in compliance with emission standards for the following at all emission sources?

    a. Suspended particulates   ___ ___ ___

    b. Sulfur dioxide   ___ ___ ___

    c. Carbon monoxide   ___ ___ ___

    d. Photochemical oxidants   ___ ___ ___

    e. Hydrocarbons   ___ ___ ___

    f. Nitrogen dioxide   ___ ___ ___

    g. Hydrogen sulfide   ___ ___ ___

20. Are there any other substances for which permit limits have been established for this facility?   ___ ___ ___

21. Have reportable excursions on emission limits occurred

    a. Six times per quarter?   ___ ___ ___

    b. 12 times per quarter?   ___ ___ ___

    c. More than 12 times per quarter?   ___ ___ ___

22. Does the facility have an up-to-date air pollution and emergency plan?   ___ ___ ___

23. Are the facility emissions and the facility location likely to contribute to interstate or international air pollution control problems?   ___ ___ ___

24. Does the facility have odorous emissions that result in complaints from the general public?   ___ ___ ___

    a. If yes, are records kept of the nature and source of the complaints?   ___ ___ ___

    b. Are these complaints routinely investigated?   ___ ___ ___

25. Are fugitive dust emissions from the facility a problem?   ___ ___ ___

	Yes	No	N/A

26. Have all permit applications been properly prepared and certified by a licensed professional engineer in the state where the facility is located?   ___ ___ ___

27. Are all meetings with, visits by, or understandings with regulatory agencies documented (i.e., approvals for variances, future actions, etc.)?   ___ ___ ___

Audit team member(s) completing questionnaires _____

_____

Location personnel providing responses _____

_____

APPENDIX E

SOLID AND HAZARDOUS WASTE MANAGEMENT AUDIT PROTOCOL

This audit protocol is intended as a guide for designing and conducting facility audits. It may require additions, revisions, or other modifications in order to meet the needs of your particular audit objectives, industrial setting, or other special circumstances.

Facility Name: _____ Date(s) of Audit: _____

Team Members Participating
in Solid and Hazardous Waste Management Audit: _____

Protocol   Protocol
Prepared By: _____ Reviewed By: _____

	Auditor(s)/ Comments	Working Paper Reference (List Page Numbers)

1. Using data supplied by the facility or provided by the audit team leader, review and examine the following:

    a. Facility layout

    b. Process flow diagrams

    c. Descriptions of known points of solid and hazardous waste generation

    d. Regulations and permits—federal, state, and local

    e. Any outstanding court orders: variances, compliance orders, administrative orders, etc.

    f. Applicable corporate policies, procedures, and standards

    g. Facility policies and procedures

    h. Instructional procedures for solid and hazardous waste handling and disposal

    i. On-site systems for handling, treating, storage, and disposal of solid and hazardous wastes

    j. Off-site systems for handling, treating, storage, transportation, and disposal of solid and hazardous wastes

2. Review the Solid and Hazardous Waste Management Questionnaire completed during the advance visit. Familiarize yourself with facility resources. (If questionnaire was not administered in advance, complete Solid and Hazardous Waste Management Questionnaire.) Make inquiries as desired to increase your understanding of the facility's responses.

	Auditor(s)/ Comments	Working Paper Reference (List Page Numbers)

3. Document in flow chart or narrative form, your understanding of the following showing responsibilities for action and record keeping for solid and hazardous wastes:

   a. Responsibility for classifying wastes

   b. Labeling

   c. Storage

   d. Shipping

   e. Sampling

   f. Maintaining manifest records

   g. Other action and record keeping activities, if any

4. At some time during the audit, tour the facility site, following the process flow diagram, for the purpose of

   a. Determining points at which solid and hazardous wastes are generated, insuring that all points are accounted for

   b. Inspecting collection, handling, and storage facilities

   c. Inspecting all treatment facilities

   d. Inspecting active disposal areas at the plant site

   e. Inspecting inactive disposal sites

   f. Inspecting PCB labeling, handling, storage, etc.

5. Waste generation

   a. Determine if points identified in (4a) actually exist. Note if unidentified points were found.

	Auditor(s)/ Comments	Working Paper Reference (List Page Numbers)

b. Verify that wastes have been tested to establish if they must be subject to RCRA rules and regulations.

c. Determine at what point waste quantities are measured and labeled and document who has responsibility.

d. Determine if wastes are sent directly to on-site or off-site treatment or disposal facilities, or if they go into storage.

e. Determine what quantities of PCB contaminated waste are generated. Determine if there are special arrangements for handling and disposal of these wastes.

f. Determine that RCRA annual federal report was filed as required.

g. Determine that any required state annual reports have been filed.

6. On-site treatment and disposal (note on facility layout)

a. Identify what is currently and has historically (during the period under review) been treated.

b. Identify treatment or disposal methods used: incineration, other thermal treatment, landfill, land treatment, physical chemical treatment, biological treatment, or underground injection.

c. Identify ultimate disposal of any residues remaining from 6(b) above.

	Auditor(s)/ Comments	Working Paper Reference (List Page Numbers)

d. Obtain operating records of all methods identified in 6(b). Inspect the records and document the information recorded for the period under review noting

   (1) If the records contain quantities and types of solid and hazardous wastes treated.

   (2) Disposal methods and disposal locations of any residues.

   (3) On-site disposal into or on land or into underground strata.

   (4) Interim status: part A permit; part B permit.

   (5) If all required construction permits were obtained.

   (6) Operational permit for treatment facilities.

   (7) If permits are currently applicable, and record dates of issuance and expiration.

   (8) If all regulatory required reports have been issued promptly in accordance with requirements. Note location where records are maintained and persons responsible.

e. For each treatment method identified in 6(b) above, verify compliance with applicable federal, state, and local rules.

	Auditor(s)/ Comments	Working Paper Reference (List Page Numbers)

7. On-site storage

   a. Record observations on layout, house-keeping, and containment.

   b. Obtain instructions for labeling and dating wastes received into storage.

   c. Inspect labels and dates on a cross-section of wastes in storage. Document agreement between inventory records and wastes in storage.

   d. If greater than 90 days, verify that a storage permit or interim status was obtained.

   e. Check quantities of RCRA-type wastes in storage against records to determine if they are consistent with permit limitations.

   f. Document the flow of records from the time of receipt of waste at storage site until removal.

   g. Determine if labels retain their integrity during handling and storage.

8. Off-site disposal

   a. Obtain and record names and addresses of all off-site contractors, including any ocean dumping.

   b. Document contractor's EPA-issued disposal number.

   c. Document that a current disposer permit exists for each disposal site.

   d. Ascertain that disposer permits are approved for the particular type of waste.

	Auditor(s)/ Comments	Working Paper Reference (List Page Numbers)

e. Document waste shipment by inspection of manifest forms. Insure that all sections of the forms have been completed. Cross-check by inspecting accounts payable files, which should indicate name or classification of waste materials, specific hazards of the wastes, and any precautions to be taken.

f. Inspect and document that shipping containers are properly marked and labeled.

g. Review contract with hauler and/or disposer, looking particularly for certificates of insurance, limitations on subcontracting, legal liabilities, etc.

h. Determine that operations at all off-site disposal locations have been inspected within last two years by company personnel.

9. If PCBs exist at this facility location, perform the following:

a. Review any action plan for insuring the removal and disposal of PCB.

b. Review that the facilities used to store PCBs designated for disposal comply with EPA rules and regulations.

c. Determine through observation that the plant is complying with marking formats established by EPA. Document your findings.

d. Review the record keeping and monitoring program for PCBs and deter-

	Auditor(s)/ Comments	Working Paper Reference (List Page Numbers)
mine that the necessary records for reporting are being maintained.		
10. Review RCRA record keeping systems (inspection logs, manifest systems, job titles, etc.).		
11. Prepared detailed list of audit findings.		

SOLID AND HAZARDOUS WASTE MANAGEMENT QUESTIONNAIRE

This audit questionnaire is intended as a guide for designing and conducting facility audits. It is intended to be used in conjunction with an audit protocol. This questionnaire may require additions, revisions, or other modifications in order to meet the needs of your particular audit objectives, industrial setting, or other special circumstances.

		Yes	No	N/A

1. Confirmation of facility as a generator

    a. Has the facility characterized all solid wastes generated to determine which are hazardous under RCRA? ___ ___ ___

    b. Does the facility produce any wastes classified as hazardous?

       Listed wastes ___ ___ ___

	Yes	No	N/A

Ignitable ___ ___ ___

Corrosive ___ ___ ___

Reactive ___ ___ ___

EP toxic ___ ___ ___

Other. Please explain. _____
_____

   c.   Does the facility have an EPA-issued generator identification number? ___ ___ ___

2.   Does the facility treat, store, or dispose of hazardous wastes on site? ___ ___ ___

   a.   Does the facility have a RCRA permit? ___ ___ ___

   b.   Does the facility have a written waste analysis plan? ___ ___ ___

   c.   Does the facility accept wastes from other facilities for storage, treatment, or disposal? ___ ___ ___

3.   Are solid and hazardous wastes treated at the facility site by any of the following methods? ___ ___ ___

   a.   Incineration ___ ___ ___

   b.   Other thermal treatment ___ ___ ___

   c.   Landfill ___ ___ ___

   d.   Land treatment, for example, land farming ___ ___ ___

   e.   Physical or chemical treatment ___ ___ ___

   f.   Biological treatment ___ ___ ___

   g.   Underground injection ___ ___ ___

4.   Does the facility accumulate or store wastes in any of the following? ___ ___ ___

   a.   Piles ___ ___ ___

   b.   Surface impoundments ___ ___ ___

		Yes	No	N/A
c.	Drums	___	___	___
d.	Tanks	___	___	___
e.	Other containers	___	___	___
5.	Are hazardous waste containers in the following condition?	___	___	___
a.	Labeled (waste type and date of accumulation)	___	___	___
b.	Compatible with wastes	___	___	___
c.	Closed	___	___	___
6.	Are storage areas inspected regularly for leaks, corrosion, or other deterioration?	___	___	___
7.	Does the facility maintain a record of the identity and location of all stored wastes?	___	___	___
8.	Is the storage area covered, diked, etc.?	___	___	___
9.	Is written inspection schedule available?	___	___	___
10.	Is inspection log kept?	___	___	___
11.	Are danger signs around active sites?	___	___	___
12.	Does the facility have written contingency plans as follows?			
a.	Available for review by EPA	___	___	___
b.	Provided to off-site emergency response team	___	___	___
13.	Does the facility have a formal emergency system?	___	___	___
a.	Emergency coordinator	___	___	___
b.	Emergency equipment	___	___	___
c.	Alarms	___	___	___

		Yes	No	N/A

14. Does the facility have a record keeping and reporting system? ___ ___ ___

    a.   For all waste management operations ___ ___ ___

          Accumulation/storage ___ ___ ___

          Treatment ___ ___ ___

          Disposal ___ ___ ___

    b.   Submit an annual report to the EPA regional administrator? ___ ___ ___

    c.   Operate an approved manifest system? ___ ___ ___

    d.   Records of incidents and reporting? ___ ___ ___

15. Is the facility required to monitor its storage, treatment, and/or disposal facilities? ___ ___ ___

    a.   Groundwater monitoring

          Has the plan been prepared by hydrologists? ___ ___ ___

          Does the plan include specification and number of wells? ___ ___ ___

          Does the plan include analysis of samples (frequency and constituents)? ___ ___ ___

    b.   Is leachate monitoring required? ___ ___ ___

16. Are hazardous wastes transported off-site? ___ ___ ___

    a.   Do transporters have EPA-issued identification numbers? ___ ___ ___

    b.   Are hazardous wastes transported by any of the following?

          Company vehicles ___ ___ ___

          Transporter vehicles ___ ___ ___

          Common carriers ___ ___ ___

    c.   Is an approved manifest system followed? ___ ___ ___

	Yes	No	N/A

d. Are Department of Transportation regulations followed?   ___ ___ ___

e. Are required reports sent to the regulatory agency?   ___ ___ ___

17. Does the facility utilize off-site facilities for the treatment, storage, or disposal of hazardous wastes?   ___ ___ ___

a. Off-site facilities for treatment   ___ ___ ___

b. Off-site facilities for storage   ___ ___ ___

c. Off-site facilities for disposal   ___ ___ ___

d. Does the off-site facility have an EPA permit?   ___ ___ ___

18. Does the off-site facility have closure and post-closure plans?   ___ ___ ___

19. Has the off-site facility owner met financial requirements of regulatory agencies or of the company?   ___ ___ ___

20. Are there any inactive waste disposal sites at the facility site?   ___ ___ ___

a. Were these reported to the Eckhardt Committee?   ___ ___ ___

b. Are any sites on EPA's National Priority List?   ___ ___ ___

c. Do they pose environmental problems?   ___ ___ ___

d. Are they monitored?   ___ ___ ___

e. Does the facility maintain records?   ___ ___ ___

21. Does the facility have a training program?   ___ ___ ___

22. Does the facility have job titles and job descriptions?   ___ ___ ___

23. Does the facility have record of training?   ___ ___ ___

		Yes	No	N/A

24. Has the facility been the subject of any RCRA enforcement actions? ___ ___ ___

25. Have off-site TSDFs been the subject of any RCRA enforcement actions? ___ ___ ___

26. Have the facility's solid and hazardous waste programs been reviewed by an outside source in the past three years? ___ ___ ___

27. Does the facility use or handle PCBs or PCB items? ___ ___ ___

    a. Are PCBs stored on-site? ___ ___ ___

    b. Are PCBs disposed of in any of the following ways?

        Incineration ___ ___ ___

        Chemical waste landfill ___ ___ ___

        High-efficiency boiler ___ ___ ___

    c. Does the facility maintain records on PCBs at the facility? ___ ___ ___

    d. Does the facility have a standard procedure for PCB decontamination? ___ ___ ___

    e. Does the facility maintain a PCB inspection program? ___ ___ ___

    f. Are PCB items appropriately labeled? ___ ___ ___

Audit team member(s) completing questionnaires _____

_____

Location personnel providing responses _____

_____

INDEX